Bob Miller's CALC II HELPER

ROBERT MILLER

Mathematics Department
City College of New York

McGraw-Hill, Inc.

New York St. Louis San Francisco Auckland Bogotá Caracas
Hamburg Lisbon London Madrid Mexico Milan Montreal
New Delhi Paris San Juan São Paulo Singapore Sydney Tokyo Toronto

This book, as everything else in my life,
is dedicated to my love, my life,
Marlene, my wife.

Bob Miller's Calc II Helper

2 3 4 5 6 7 8 9 10 11 12 13 14 15 16 17 18 19 20 SHP SHP 9 2 1

ISBN 0-07-042258-3

Sponsoring Editor, David Beckwith
Production Supervisor, Anita Kann
Editing Supervisors, Meg Tobin, Maureen Walker, Patty Andrews
Designer, Wanda Siedlecka
Chapter opening drawing by Leslie Cober;
Cover photo by Zygmunt Malinowski.

Library of Congress Cataloging-in-Publication Data

Miller, Robert A.
Bob Miller's calc II helper / Robert Miller.
p. cm.
Includes index.
ISBN 0-07-042258-3
1. Calculus. I. Title. II. Title: Beginning calc helper.
QA303.M6815 1991
515--dc20 90-13264
CIP

ABOUT BOB MILLER . . . IN HIS OWN WORDS.

I received my B.S. and M.S. in math from Brooklyn Poly (now Polytechnic Institute of New York). After teaching my first class there as a substitute for a full professor, one student, upon leaving the room, told another that at least now we have someone who could teach the stuff. I was forever hooked on teaching. Since then I have taught at Westfield State College, Westfield, Massachusetts, Rutgers, and the City College of New York where I've been forever (actually 22 1/2 years). No matter how badly I feel, I always feel great after I start teaching. I especially love to teach precalc and calc courses and am always delighted when a student tells me that he or she has always hated math before and never could learn it but taking a class with me has made math understandable and even enjoyable. I've got a wonderful wife and two great children. I enjoy golf, bowling, bridge, and crossword puzzles, but to me teaching is the greatest joy in the world.

TO THE STUDENT

This book was written for you: not your teacher, not your next-door neighbor, not for anyone but you. I have tried to make the examples and explanations as clear as I can. However, as much as I hate to admit it, I am not perfect. If you find something that is unclear or should be added to this book, please let me know. If you want a response, or I can help you, your class, or your school, in any precalculus or calculus subject, please let me know, but address your comments c/o McGraw-Hill, Inc., Schaum Division.

If you make a suggestion on how to teach one of these topics better and you are the first and I use it, I will give you credit for it in the next edition.

Please be patient on responses. I am hoping the book is so good that millions of you will write. I will answer.

Now, enjoy the book and learn.

ROBERT MILLER

CONTENTS

1. LOGARITHMS

Most of you, at this point in your mathematics, have not seen logs for at least a year, many a lot more. The normal high school course emphasizes the wrong areas. You spend most of the time doing endless calculations, none of which you need here. By the year 2000, students will do almost no log calculations due to calculators. In case you feel tortured, just remember you only spent weeks on log calculations. I spent months!!!

THE BASIC LAWS OF LOGS

1. Definition $\log_b x = y$ (log of x to the base b is y) if $b^y = x$. $\log_5 25 = 2$ because $5^2 = 25$.
2. What can the base b be? It can't be negative, such as -2, since $(-2)^{1/2}$ is imaginary. It can't be 0, since 0^n is either equal to 0 if n is positive or undefined if n is 0 or negative. b also can't be 1 since 1^n always $= 1$. Therefore b can be any positive number except 1.

   Note: The base can be $2^{1/2}$, but it won't do you any good because there are no $2^{1/2}$ tables. The 2 most common bases are 10, because we have 10 fingers, and e, a number that occurs a lot in mathematics starting now.

   A. e equals approximately 2.7.

   B. e more exactly? On a calculator press 1, inv, ln.

   C. $\log = \log_{10}$

   D. $\ln = \log_e$ (ln is the natural logarithm).

3. Range of logs, y: A log, y, is an exponent and exponents can be positive, negative, and zero. The range is all real numbers.

4. Domain, x values: Since the base is positive, whether the exponent is positive, zero, or negative, the answer is positive. The domain therefore is positive numbers.

   Note: In order to avoid getting too technical, most books write $\log|x|$, thereby excluding only $x = 0$.

5. $\log_b x + \log_b y = \log_b xy$. $\log 2 + \log 3 = \log 6$.

6. $\log_b x - \log_b y = \log_b (x/y)$. $\log 7 - \log 3 = \log (7/3)$.

7. $\log_b x^p = p \log_b x$. $\ln 6^7 = 7 \ln 6$ is OK.

Note: Laws 5,6,7 are most important. If you can simplify using these laws, about half the battle (the easy half) is done.

EXAMPLE 1 Write the following as simpler logs with no exponents: $\ln\left(\frac{a^4b^5}{c^6\sqrt{d}}\right)$.

$4 \ln a + 5 \ln b - 6 \ln c - \frac{1}{2} \ln d$

8. $\log_b b = 1$ since $b^1 = b$. $\log_7 7 = 1$. $\ln e = 1$. $\log 10 = 1$.

9. $\log_b 1 = 0$ since $b^0 = 1$. $\log_8 1 = \log 1 = \ln 1 = 0$.

10. Log is a 1-1 function. This means if $\log c = \log d$, $c = d$.

    Note: Not everything is 1-1. If $x^2 = y^2$, $x = \pm y$.

11. Log is an increasing function. If $m < n$, then $\log m < \log n$.

12. $\lim_{x\to\infty} (1 + 1/x)^x = e$

13. $b^{\log_b x}$ is a weird way of writing x. $e^{\ln x} = x$.

14. $\log_b b^x = x$. $\ln e^x = x$.

15. $\log_d c = \frac{\log_b c}{\log_b d}$. $\log_d c = \frac{\ln c}{\ln d}$. $\log_2 5 = \frac{\ln 5}{\ln 2}$.

You should be able to solve the following kinds of log equations:

EXAMPLE 2 Solve for x: $4 \cdot 3^{x+2} = 28$.

$3^{x+2} = 7$ **Divide by 4; isolate exponent**

$$(x + 2) \log 3 = \log 7$$

Take logs. It now becomes an elementary algebra equation, which we solve for x, using same technique as in implicit differentiation section of Calc I

$$x = \frac{\log 7 - 2 \log 3}{\log 3}$$

16. $a^x = e^{x \ln a}$. Also $x^x = e^{x \ln x}$ and $x^{\sin x} = e^{\sin x \ln x}$.

EXAMPLE 3 $4^{2x+5} = 9^{3x-7}$

$$(2x + 5) \log 4 = (3x - 7) \log 9$$

Using the same algebraic tricks we get

$$x = \frac{-7 \log 9 - 5 \log 4}{2 \log 4 - 3 \log 9} \quad \text{or} \quad \frac{7 \log 9 + 5 \log 4}{3 \log 9 - 2 \log 4}$$

Eliminate excess minus signs

All this should be known about logs before the calculus. Now we are ready to get serious.

17. Major theorem. Given $f(x) = \ln x$; then $f'(x) = 1/x$.

Proof (worth looking at):

$$\lim_{\Delta x \to 0} \frac{f(x + \Delta x) - f(x)}{\Delta x} = \frac{\ln (x + \Delta x) - \ln x}{\Delta x}$$

Definition of derivative

$$= \lim_{\Delta x \to 0} \frac{\ln \left(\frac{x + \Delta x}{x} \right)}{\Delta x}$$

Rule 6

$$= \lim_{\Delta x \to 0} \frac{x \ln \left(\frac{x + \Delta x}{x} \right)}{x \Delta x}$$

Normal trick 1—multiply by $1 = x/x$

$$= \lim_{\Delta x \to 0} \frac{\frac{x}{\Delta x} \ln \left(1 + \frac{\Delta x}{x} \right)}{x}$$

Algebra

$$= \lim_{\Delta x \to 0} \frac{\ln \left(1 + \frac{\Delta x}{x} \right)^{x/\Delta x}}{x}$$

Rule 7

$$= \lim_{w \to \infty} \frac{(1 + 1/w)^w}{x}$$

Use trick 2, $\frac{x}{\Delta x} = w$
As $\Delta x \to 0$, $w \to \infty$

$$= \frac{\ln e}{x} \qquad \textbf{Rule 12}$$

$$= 1/x \qquad \textbf{Rule 8}$$

This theorem is important since it has a lot of log rules together with 2 normal math tricks.

The theorem gives us the following result:

$$\int \frac{1}{x}\, dx = \ln |x| + C$$

2. DERIVATIVES OF e^x, a^x, LOGS, TRIG FUNCTIONS, ETC. AND ETC.

We will now take derivatives involving $\ln x$, e^x, a^x, $f(x)^{g(x)}$, trig functions, and inverse trig functions.

EXAMPLE 1 $y = \ln(x^2 + 5x + 7)$

Let $u = x^2 + 5x + 7$. Then $y = \ln u$. So $dy/dx = (dy/du)(du/dx) = (1/u)(2x + 5) = (2x + 5)/(x^2 + 5x + 7)$.

Notice, taking derivatives of logs is not difficult. However, you do not want to actually substitute $u = x^2 + 5x + 7$. You must do that in your head. If $y = \ln u$, do $y' = (1/u)(du/dx)$ in your head!

EXAMPLE 2 $y = \ln \dfrac{(x^2 + 7)^9(x + 3)}{x^6}$

The simplest way to do this is use laws 3,4,5 of the preceding chapter and simplify the expression before we take the derivative. So $y = 9\ln(x^2 + 7) + \ln(x + 3) - 6\ln x$. Therefore

$$y' = \frac{9(2x)}{x^2 + 7} + \frac{1}{x + 3} - \frac{6}{x}$$

Remember to simplify by multiplying $9(2x) = 18x$

EXAMPLE 3 $y = \log_2 x$

Using law 15, $y = \dfrac{\ln x}{\ln 2}$, where $\ln 2$ is a number (a constant). Therefore

$$y' = (1/\ln 2)(\ln x)' = 1/(x \ln 2)$$

LAW 18 $y = e^u$. $y' = e^u(du/dx)$. If $y = e$ to the power u, where $u =$ a function of x, the derivative is the original function untouched times the derivative of the exponent.

EXAMPLE 4 $y = e^{\sin x}$

$y' = e^{\sin x} (\sin x)' = e^{\sin x} (\cos x)$

LAW 19 $y = a^u$. $y' = a^u \ln a(du/dx)$. If $y = a^u$, the derivative is a^u (the original function untouched) times the log of the base times the derivative of the exponent.

EXAMPLE 5 $y = 7^{x^2}$

$y' = (7^{x^2})(\ln 7)(2x)$

$= $ (whole function) (ln of base) (deriv. of exponent)

Let us, for completeness, recall the trig derivatives, and do one longish chain rule

LAW 20

A. $y = \sin x$. $y' = \cos x$.

B. $y = \cos x$. $y' = -\sin x$.

C. $y = \tan x$. $y' = \sec^2 x$.

D. $y = \cot x$. $y' = -\csc^2 x$.

E. $y = \sec x$. $y' = \tan x \sec x$.

F. $y = \csc x$. $y' = -\cot x \csc x$.

EXAMPLE 6 $y = \tan^6 (4x^2 + 3x + 7)$

Since this is a function of a function, we must use the extended chain rule.

Let $u = \tan (4x^2 + 3x + 7)$. $y = u^6$ and $dy/du = 6u^5$. Let $v = 4x^2 + 3x + 7$. $u = \tan v$. $du/dv = \sec^2 v$ and $dv/dx = 8x + 3$. So

$dy/dx =$	(dy/du) times	(du/dv) times	(dv/dx)
$=$	$6u^5$ times	$\sec^2 v$ times	$(8x + 3)$
$=$	$[6 \tan^5 (4x^2 + 3x + 7)]$	$[\sec^2 (4x^2 + 3x + 7)]$	$(8x + 3)$
	Power rule—leave trig function and crazy angle untouched	Derivative of trig function—leave crazy angle untouched	Derivative of crazy angle

You should be able to do this without substituting for u and v. It really is not that difficult with a little practice.

LAW 21

A. $y = \sin^{-1} u$. $y' = \dfrac{1}{(1-u^2)^{1/2}}\, du/dx \quad |u| < 1.$

B. $y = \tan^{-1} u$. $y' = \dfrac{1}{1+u^2}\, du/dx.$

C. $y = \sec^{-1} u$. $y' = \dfrac{1}{u(u^2-1)^{1/2}}\, du/dx \quad |u| > 1.$

EXAMPLE 7 $y = \tan^{-1}(e^{5x})$

$u = e^{5x}$. $u' = e^{5x}\,5$.

$$y' = \frac{1u'}{1+u^2} = \frac{5e^{5x}}{1+(e^{5x})^2} = \frac{5e^{5x}}{1+e^{10x}}$$

EXAMPLE 8 $y = (\tan^{-1} x)/(1+x^2)$

Use the quotient rule.

$$y' = \frac{(1+x^2)[1/(1+x^2)] - (\tan^{-1} x)(2x)}{(1+x^2)^2} = \frac{1 - 2x\tan^{-1} x}{(1+x^2)^2}$$

I like this next one. I don't know why, but I do.

EXAMPLE 9 $y = \sin^{-1}(\cos^2 x)$

$u = \cos^2 x$. $u' = 2(\cos x)(-\sin x)$.

$$y' = \frac{u'}{(1-u^2)^{1/2}} = \frac{-2\cos x \sin x}{(1-\cos^4 x)^{1/2}}$$

EXAMPLE 10 $y = \sec^{-1}(x^2+1)$

$u = x^2 + 1$. $u' = 2x$.

$$y' = \frac{u'}{u(u^2-1)^{1/2}} = \frac{2x}{(x^2+1)[(x^2+1)^2-1]^{1/2}}$$

If $y = f(x)^{g(x)}$, we take logs of both sides and differentiate implicitly (if you've forgotten differentiation implicitly, see my Calc I).

EXAMPLE 11 $y = x^{\sin x}$

$$\ln y = \sin x \ln x$$

ln y = sin x ln x

$$(1/y)(dy/dx) = (\sin x)(1/x) + (\ln x)(\cos x)$$

$$dy/dx = y[(\sin x)/x + (\ln x)(\cos x)]$$

$$= x^{\sin x}[(\sin x)/x + (\ln x)(\cos x)]$$

EXAMPLE 11, alternative method $y = x^{\sin x} = e^{\sin x \ln x}$

$$y' = e^{\sin x \ln x}(\sin x \ln x)'$$

$$= e^{\sin x \ln x}[(\sin x)/x + (\ln x)(\cos x)]$$

$$= x^{\sin x} \sin [(x/x) + (\ln x)(\cos x)]$$

EXAMPLE 12 $y = \dfrac{(x^3+1)^7(\sin^6 x)(x^2+4)^9}{e^{7x}x^8}$

To take derivatives without logs is long and leads to errors. However taking logs first and differentiating implicitly makes things much shorter and easier.

$$\ln y = 7 \ln (x^3 + 1) + 6 \ln \sin x + 9 \ln (x^2 + 4) - 7x - 8 \ln x$$

$$(1/y)\, dy/dx = \frac{7(3x^2)}{x^3+1} + \frac{6(\cos x)}{\sin x} + \frac{9(2x)}{x^2+4} - 7 - \frac{8}{x}$$

$$\frac{dy}{dx} = y\left[\frac{21x^2}{x^3+1} + 6 \cot x + \frac{18x}{x^2+4} - 7 - \frac{8}{x}\right]$$

This looks neat. But remember what y really is!! But it is pretty short!

3. SHORTER INTEGRALS

In most schools, the largest part of the second semester of a three-term calc sequence involves integrals. It usually covers more than 50% of this course. It is essential to learn these shorter ones as perfectly as possible so that Chapter 6 will not be overwhelming. Also it is impossible to put every pertinent example in without making the book too long. The purpose of this book is to give you enough examples so that you can do the rest by yourself. If you think an example should be added, write me.

Rule 1 $\int \frac{f'(x)\,dx}{f(x)} = \ln|f(x)| + C$

One of the first new things we look for is that the numerator is the derivative of the denominator. This gives us a ln for an answer.

EXAMPLE 1

$$\int \frac{x\,dx}{5x^2-7}$$ **Let $u = 5x^2 - 7$. $du = 10x\,dx$**

$$= (1/10)\int \frac{10x\,dx}{5x^2-7} = (1/10)\int \frac{du}{u}$$
$$= (1/10)\ln|u| + C = (1/10)\ln|5x^2-7| + C$$

EXAMPLE 2

$$\int \frac{\cos x\,dx}{1+\sin x}$$ **$u = 1 + \sin x$. $du = \cos x\,dx$**

$$= \int \frac{du}{u} = \ln|u| + C = \ln(1+\sin x) + C$$

Exclude $x = 3\pi/2$, etc. Then $\sin x > -1$, so abs. value not needed in answer

EXAMPLE 3

$$\int \frac{6\,dx}{x^{1/2}(x^{1/2}+3)}$$

This one looks kind of weird. Sometimes we just have to try something. Let $u = x^{1/2} + 3$ (note $u = x^{1/2}$ will also work).
$du = \frac{1}{2}x^{-1/2}\,dx$ so
$dx = 2x^{1/2}\,du$

$$= \int \frac{6(2x^{1/2}\,du)}{x^{1/2}u} = \int \frac{12\,du}{u}$$

$$= 12 \ln u + C = 12 \ln (x^{1/2} + 3) + C$$

Let's try a definite integral.

EXAMPLE 4

$$\int_{e}^{e^e} \frac{dx}{x \ln x}$$

$u = \ln x.\ du = (1/x)\,dx$

$$= \int (1/u)\,du = \ln u = \ln (\ln x) \Big[_{e}^{e^e}$$

$$= \ln (\ln e^e) - \ln (\ln e) = \ln (e) - \ln 1 = 1 - 0 = 1$$

EXAMPLE 5

$$\int \left(\frac{x^2 - 4x + 3}{x - 2}\right) dx$$

We need to long divide since the degree of the top is greater than the bottom:

$$= \int \left[x - 2\ \frac{-1}{x-2}\right] dx$$

$$= \frac{x^2}{2} - 2x - \ln |x - 2| + C$$

$$\begin{array}{r l}
 & x \ -2\ +\ -1/(x-2) \\
x - 2\,) & \overline{x^2 - 4x + 3} \\
 & \underline{x^2 + 2x} \\
 & -2x + 3 \\
 & \underline{-2x - 4} \\
 & -1
\end{array}$$

EXAMPLE 6

$$\int \frac{2x\,dx}{(x^2+3)^2}$$

$u = x^2 + 3.\ du = 2x\,dx.$ WARNING! This is NOT a logarithm!! The exponent on the bottom is 2!!! It must be 1 to be a log!!!

$$= \int \frac{du}{u^2} = \int u^{-2}\,du = \frac{-1}{u} + C = \frac{-1}{x^2 + 3} + C$$

TRIG INTEGRALS

Rule(s) 2

A. $\int \sin x\,dx = -\cos x + C$

B. $\int \cos x\,dx = \sin x + C$

C. $\int \sec^2 x\,dx = \tan x + C$

D. $\int \tan x\,dx = -\ln|\cos x|$ or $\ln|\sec x| + C$

E. $\int \csc^2 x\,dx = -\cot x + C$

F. $\int \cot x\,dx = \ln|\sin x| + C$

G. $\int \tan x \sec x\,dx = \sec x + C$

H. $\int \sec x\,dx = \ln|\tan x + \sec x| + C$

I. $\int \csc x \cot x\,dx = -\csc x + C$

J. $\int \csc x\,dx = -\ln|\cot x + \csc x| + C$ or
$= \ln|\csc x - \cot x| + C$

YOU MUST KNOW THESE INTEGRALS PERFECTLY!!!!!!!!!

EXAMPLE 7

$\int \cot 4x\,dx$ **$u = 4x$. $du = \frac{1}{4}\,dx$**

$$= \tfrac{1}{4}\int \cot u\,du = \tfrac{1}{4}\ln|\sin u| + C = \tfrac{1}{4}\ln|\sin(4x)| + C$$

Note: Whenever you have the integral of one of these trig functions and there is a constant multiplying the angle, you must, by sight, integrate this without letting u equal the angle. Otherwise integrals in Chapter 6 will take forever.

EXAMPLE 8

$\int x^2 \cos(1 - 3x^3)\,dx$ **This is the crazy angle substitution. u = crazy angle = $1 - 3x^3$. $du = -9x^2\,dx$**

$$= -(1/9)\int (-9x^2)\cos(1-3x^3)\,dx$$

$$= (-1/9)\int \cos u\,du = (-1/9)\sin u + C$$

$$= (-1/9)\sin(1-3x^3) + C$$

EXAMPLE 9 $\int \tan 2x \sec^2 2x\,dx$ **$u = \tan 2x$. $du = 2\sec^2 2x\,dx$**

$$= \tfrac{1}{2}\int u\,du = \tfrac{1}{2}u^2/2 + C = \tfrac{1}{4}\tan^2 2x + C$$

EXAMPLE 10 $\int \frac{1+\sin x\,dx}{\cos^2 x}$ **This one requires splitting the integrand into 2 fractions and uses identities**

$$= \int \left(\frac{1}{\cos^2 x} + \frac{\sin x}{\cos^2 x}\right) dx$$

$$= \int (\sec^2 x + \tan x \sec x)\,dx = \tan x + \sec x + C$$

Easy one if (a BIG if) you know your identities and trig integrals.

EXPONENTIAL INTEGRALS

Rule 3 $\int e^{bx}\,dx = \frac{e^{bx}}{b} + C$. Know this perfectly by sight!

EXAMPLE 11 $\int e^{5x}\,dx = (e^{5x})/5 + C$

Rule 4 $\int a^{bx}\,dx = \frac{a^{bx}}{b \ln a} + C$. Know this one perfectly also!

EXAMPLE 12 $\int 7^{5x}\,dx = \frac{7^{5x}}{5 \ln 7} + C$

EXAMPLE 13 $\int \frac{e^{7/x}}{x^2}\,dx$ **Crazy exponent substitutions. $u = 7/x$. $du = -7/x^2\,dx$**

$$= \int -\frac{e^u}{7}\,du = -\frac{e^u}{7} + C = -\frac{e^{7/x}}{7} + C$$

EXAMPLE 14

$$\int \frac{7^{\ln x}}{x}\,dx$$ **Crazy exponent. u = ln x. du = (1/x) dx**

$$= \int 7^u\,du = \frac{7^u}{\ln 7} + C = \frac{7^{\ln x}}{\ln 7} + C$$

EXAMPLE 15

$$\int \frac{e^{\sin x}\,dx}{\sec x}$$ **Crazy exponent (only real choice) plus trig identity. u = sin x. du = cos x dx = dx/sec x**

$$= \int e^u\,du = e^u + C = e^{\sin x} + C$$

EXAMPLE 16

$$\int \frac{e^{4x}\,dx}{(e^{4x}+1)^2}$$ **One of my favorites. This one looks exactly like the next one,* but is really different. $u = e^{4x} + 1$. $du = 4e^{4x}\,dx$**

$$= \frac{1}{4}\int \frac{4e^{4x}\,dx}{(e^{4x}+1)^2} = \frac{1}{4}\int \frac{du}{u^2} = \frac{-1}{4u} + C = \frac{-1}{e^{4x}+1} + C$$

INVERSE TRIG FUNCTIONS

This part is the last of the basic integrals that you must know by sight. In some schools, all 6 inverse trig functions must be known; in some, 3; and in some, like in my school, 2. We will do three—arc sin, arc tan, arc sec.

Rule(s) 5

$$\int \frac{1\,dx}{(1-x^2)^{1/2}} = \sin^{-1} x + C. \qquad \int \frac{1\,dx}{(a^2-x^2)^{1/2}} = \sin^{-1}(x/a) + C.$$

$$\int \frac{1\,dx}{1+x^2} = \tan^{-1} x + C. \qquad \int \frac{1\,dx}{a^2+x^2} = (1/a)\tan^{-1}(x/a) + C.$$

$$\int \frac{1\,dx}{x(x^2-1)^{1/2}} = \sec^{-1} x + C. \qquad \int \frac{1\,dx}{x(x^2-a^2)^{1/2}} = (1/a)\sec^{-1}(x/a) + C.$$

Memorize these also!

***EXAMPLE 16A** **(the next one I left out)**

$$\int \frac{(e^{4x}+1)^2}{e^{4x}}\,dx$$ **This is different**

$$= \int \frac{e^{8x} + 2e^{4x} + 1}{e^{4x}}\,dx = \int e^{4x} + 2 + e^{-4x}\,dx = \frac{e^{4x}}{4} + 2x - \frac{e^{-4x}}{4} + C$$

These integrals are not long, but you must study because there are a lot of differences.

You will be able to identify these integrals by sight with practice. As for me, I know arc sin and arc tan very well, because I've practiced them. However in all the years I've taught, no one has ever required arc sec; so I have to struggle since I need practice also. Practice is what is needed!

EXAMPLE 17

$$\int \frac{e^{3x}\,dx}{(1-e^{6x})^{1/2}}$$

You must see this is an arc sin. $u = e^{3x}$. $u^2 = (e^{3x})^2 = e^{6x}$. $du = 3e^{3x}\,dx$. $du/3 = e^{3x}\,dx$

$$= \int \frac{(1/3)\,du}{(1-u^2)^{1/2}} = (1/3)\text{ arc sin } u + C = (1/3)\text{ arc sin }(e^{3x}) + C$$

EXAMPLE 18

$$\int \frac{x^5\,dx}{7+x^{12}}$$

You must see this is an arc tan. $u = x^6$. $u^2 = (x^6)^2 = x^{12}$. $du = 6x^5\,dx$. $a = 7^{1/2}$ since $a^2 = 7$

$$= (1/6)\int \frac{du}{7+u^2} = \frac{1}{6\cdot 7^{1/2}}\tan^{-1}(x^6/7^{1/2}) + C$$

EXAMPLE 19

$$\int \frac{dx}{x(x^4-11)^{1/2}}$$

This is harder to tell. It is an arc sec with $u = x^2$, $du = 2x\,dx$. Multiply top and bottom by $2x$.

$$= \frac{1}{2}\int \frac{2x\,dx}{x^2(x^4-11)^{1/2}} = \frac{1}{2}\int \frac{du}{u(u^2-11)^{1/2}}$$

$a = 11^{1/2}$

$$= \frac{1}{2}(1/11^{1/2})\text{ arc sec }(u/11^{1/2}) + C = \frac{\text{arc sec }(x^2/11^{1/2})}{2\cdot 11^{1/2}} + C$$

You can all do it if you concentrate and practice a little.

WARNING!!! BEWARE! DANGER! Now that you know these 3, be careful of those that look something like that but are not arcs.

EXAMPLE 20

$$\int \frac{x\,dx}{x^2+4}$$

This looks like an arc tan, but a u-substitution will give us a log. $u = x^2+4$. $du = 2x\,dx$. $du/2 = x\,dx$

$$= \tfrac{1}{2}\int du/u = \tfrac{1}{2}\ln u + C = \tfrac{1}{2}\ln(x^2+4) + C$$

EXAMPLE 21

$$\int \frac{x\,dx}{(1-x^2)^{1/2}}$$

Not an arc sin. $u = (1-x^2)$. $du = -2x\,dx$

$$= -\tfrac{1}{2}\int u^{-1/2}\,du = -u^{1/2} + C = -(1-x^2)^{1/2} + C$$

4. EXPONENTIAL GROWTH AND DECAY

In every book on calculus, there is a little on differential equations, that is, an equation with derivatives. Usually 1 chapter is devoted to this topic, which is almost never used. Parts of 1 or 2 other chapters may have differential equations in them. This topic is almost universally covered by all courses.

EXAMPLE The current rate of change of Grumbium is proportional to the current amount. (a) If 10 pounds of Grumbium becomes 50 pounds in 7 hours, when will there be 90 pounds, and (b) how much in 20 hours?

The equation is $dG/dt = kG$. One way to solve it is by separation of variables. Dividing both sides by G and multiplying both by dt, we get $dG/G = k\,dt$. Integrating we get $\ln G = kt + C$.

Let $c = \ln G_0$, Grumbium at time $t = 0$. Then setting $t = 0$ in our equation, we get $\ln G_0 = C$; so $\ln G = kt + \ln G_0$ or $\ln G - \ln G_0 = kt$. Using a property of logs we get $\ln(G/G_0) = kt$. Using the definition of ln we get $G/G_0 = e^{kt}$ or $G = G_0e^{kt}$. This is the equation you are usually allowed to start from, but we show it because in differential equations, you must be able to do this.

$G = G_0e^{kt}$. $G_0 = 10$, $G = 50$, $t = 7$. We can get k. $50 = 10e^{7k}$. Divide by 10 and take ln s; we get $\ln 5 = 7k$ or $k = (\ln 5)/7$. So . . . $G = 10e^{(\ln 5)t/7}$ or $G = 10(5^{t/7})$.

Now we are ready for part (a): $90 = 10\ 5^{t/7}$. Dividing by 10 and taking ln s, we get . . . $\ln 9 = (t/7)\ln 5$ or $t = (7\ln 9)/\ln 5$ hours. You usually can leave the answer in terms of ln s.

The easier part (b): $G = 10(5^{t/7})$. $t = 20$. So $G = 10(5^{20/7})$ pounds. This is the exact answer. A calculator will give a more usable form.

5. WHAT YOU SHOULD KNOW FROM BEFORE TO DO THE NEXT

We have now come to the part of the book that requires you to work harder than perhaps at any time in the entire calculus sequence. We are about to embark on learning new, long integration techniques. Since the product and quotient rules do not hold for integrals, we are forced to learn many techniques, most of which are long.

In order to make these integrals shorter, we are listing the previous facts that if you have properly learned, will make the following chapter much easier.

1. Definition of the 6 trig functions
2. The values of the 6 trig functions for multiples of 30, 45, 60, 90 unless your instructor allows you to cheat and use calculators
3. The derivatives of the 6 trig functions
4. For the last time, the following identities:

   A. $\sin x \csc x = 1$ B. $\cos x \sec x = 1$

   C. $\tan x \cot x = 1$ D. $\tan x = \sin x/\cos x$

   E. $\cot x = \cos x/\sin x$ F. $\sin^2 x + \cos^2 x = 1$

   G. $1 + \tan^2 x = \sec^2 x$ H. $1 + \cot^2 x = \csc^2 x$

   I. $\sin 2x = 2 \sin x \cos x$

Note: It is really of interest to note that you really don't need to know $\cos 2x$ as we will see shortly.

   J. $\sin^2 x = \dfrac{1 - \cos 2x}{2}$ K. $\cos^2 x = \dfrac{1 + \cos 2x}{2}$

5. The beginning integrals

   A. Integral of x^n $n \neq -1$

B. Multiplying out

C. Dividing out

D. u-substitution in a parenthesis

E. u-sub. for a crazy angle

F. u-sub. for a crazy exponent

6. Trig integrals

A. $\int \sin ax\,dx = (1/a)(-\cos ax) + C$

B. $\int \cos ax\,dx = (1/a)(\sin ax) + C$

C. $\int \sec^2 ax\,dx = (1/a)(\tan ax) + C$

D. $\int \csc^2 ax\,dx = (1/a)(-\cot ax) + C$

E. $\int \tan ax \sec ax\,dx = (1/a) \sec ax + C$

F. $\int \cot ax \csc ax\,dx = -(1/a) \csc ax + C$

G. $\int \tan ax\,dx = -(1/a) \ln|\cos ax| + C$ or $(1/a) \ln|\sec ax| + C$

H. $\int \cot ax\,dx = (1/a) \ln|\sin ax| + C$

I. $\int \sec ax\,dx = (1/a) \ln|\sec ax + \tan ax| + C$

J. $\int \csc ax\,dx = -(1/a) \ln|\csc ax + \cot ax| + C$ or $= (1/a) \ln|\csc ax - \cot ax| + C$

7. Definition, certain values involving multiples of 30, 45, 60, and 90 degrees, and derivatives of arc sin, arc tan, and arc sec

8. Inverse trig integrals

A. $\int \frac{dx}{(a^2 - x^2)^{1/2}} = \text{arc sin}\,(x/a) + C$

B. $\int \frac{dx}{a^2 + x^2} = (1/a)\,\text{arc tan}\,(x/a) + C$

C. $\int \frac{dx}{x(x^2 - a^2)^{1/2}} = (1/a)\,\text{arc sec}\,(x/a) + C$

9. Other integrals you should know

A. $\int 1/x\,dx = \ln|x| + C$

B. $\int f'(x)/f(x)\,dx = \ln|f(x)| + C$

C. $\int \sinh ax\,dx = (1/a)\cosh ax + C$

D. $\int \cosh ax\,dx = (1/a)\sinh ax + C$

E. $\int e^{ax}\,dx = (1/a)e^{ax} + C$

F. $\int b^{ax}\,dx = \dfrac{(\ln b)b^{ax}}{a \ln b} + C$

It is quite a list, but as you will see, all are needed.

6. INTEGRATION BY PARTS

As you will see there is very little theory in this chapter—only *HARD* work.

Integration by parts comes from the product rule for differentials, which is the same as the product rule for derivatives.

Let u and v be functions of x.

$$d(uv) = u\,dv + v\,du \qquad \text{or} \qquad u\,dv = d(uv) - v\,du$$

Integrating we get

$$\int u\,dv = \int d(uv) - \int v\,du \qquad \text{or}$$

$$\int u\,dv = uv - \int v\,du$$

What have we done? In the first integral, we have the function u and the differential of v. In the last integral we have the differential of u and the function v. By reversing the roles of u and v, we hope to either have a very easy second integral or to be allowed to proceed more easily to an answer.

EXAMPLE 1 $\int xe^{3x}\,dx$

If a polynomial multiplies e^{ax}, sin ax, and cos ax, we always let u = polynomial and $dv = e^{ax}\,dx$, sin ax dx, or cos ax dx. In this example

$$u = x \qquad dv = e^{3x}\,dx \qquad v = \frac{e^{3x}}{3} \qquad du = 1\,dx$$

$$\int \overset{u}{x}\,\overset{du}{e^{3x}\,dx} = \overset{u}{x}\,\overset{v}{\frac{e^{3x}}{3}} - \int \overset{v}{\frac{e^{3x}}{3}}\,\overset{du}{dx}$$

$$= \frac{xe^{3x}}{3} - \frac{e^{3x}}{9} + C$$

EXAMPLE 2 $\int x^4 e^{3x}\,dx$

We must let u = polynomial and $dv = e^{3x}\,dx$ 4 times!! However if you observe the pattern, in time you may be able to do this in your head. Yes, I mean you. Signs alternate, polynomial gets the derivative taken, a 3 is multiplied on the bottom each time, and e^{3x} multiplies each term.

$$\text{Answer will be } \left(\frac{x^4}{3} - \frac{4x^3}{9} + \frac{12x^2}{27} - \frac{24x}{81} + \frac{24}{243}\right)e^{3x} + C.$$

Next we will consider integrating the arc sin, arc tan, and ln. If you had never seen them before, you probably would never guess that all are done by integration by parts since there appears to be only 1 function. However, mathematicians, being clever little devils, invented a second function so that all 3 of these integrals are rather easily done.

EXAMPLE 3 $\int \sin^{-1} dx$

$$\text{Let } u = \sin^{-1} x \qquad du = \frac{1\,dx}{(1-x^2)^{1/2}}$$

$$\text{Let } dv = 1\,dx!!! \qquad v = x$$

$$\int \sin^{-1} x\,dx = \int \underset{u}{\sin^{-1} x}\ \underset{dv}{dx} = \underset{v}{x}\ \underset{u}{\sin^{-1} x} - \int \underset{v}{x}\ \underset{du}{\frac{1\,dx}{(1-x^2)^{1/2}}}$$

$$= x \sin^{-1} x + \int \frac{-x\,dx}{(1-x^2)^{1/2}} \qquad w = 1 - x^2;\ dw = -2x\,dx$$

$$= x \sin^{-1} x + \frac{1}{2}\int \frac{-2x\,dx}{(1-x^2)^{1/2}}$$

$$= x \sin^{-1} x + \frac{1}{2}\int w^{-1/2}\,dw$$

$$= x \sin^{-1} x + \frac{1}{2}\,\frac{w^{1/2}}{\frac{1}{2}} + C$$

$$= x \sin^{-1} x + w^{1/2} + C$$

$$= x \sin^{-1} x + (1-x^2)^{1/2} + C$$

We will now do a more complicated problem, $\int e^{5x} \cos 3x\,dx$. Based on what we did before, we can take either function as u and the rest as dv. It turns out both will work. However the problem is not quite so easy, as we will see. Being a glutton for punishment, I will do it both ways to show that the problem can be done 2 ways.

EXAMPLE 4 $\int e^{5x} \cos 3x\, dx$. $u = e^{5x}$. $du = 5e^{5x}\, dx$. $dv = \cos 3x$. $v = (\sin 3x)/3$.

$$\int \overset{u}{e^{5x}}\ \overset{dv}{\cos 3x\ dx} = \frac{\overset{u}{e^{5x}}\ \overset{v}{\sin 3x}}{3} - \int \overset{v}{\frac{\sin 3x}{3}}\ \overset{du}{(5e^{5x})}\, dx$$

$$= \frac{e^{5x} \sin 3x}{3} - (5/3) \int e^{5x} \sin 3x\, dx$$

At this point you might say, "This doesn't do anything for us." You'd be right. Let us do it again. We let $U = e^{5x}$ because if we reversed, we would wind up with the original integral and would have accomplished nothing. $dV = \sin 3x\, dx$. $V = (-\cos 3x)/3$.

$$\int e^{5x} \cos 3x\, dx = \frac{e^{5x} \sin 3x}{3} - (5/3) \int \overset{U}{e^{5x}}\ \overset{dV}{\sin 3x\ dx}$$

$$= \frac{e^{5x} \sin 3x}{3} - (5/3)\ \overset{U}{e^{5x}}\ \overset{V}{\frac{(-\cos 3x)}{3}}$$

$$-{}^{*}\ \frac{5}{3} \int \overset{dU}{5e^{5x}}\ \overset{V}{\frac{\cos 3x}{3}}\, dx$$

$$\int e^{5x} \cos 3x\, dx = \frac{e^{5x} \sin 3x}{3} + \frac{5e^{5x} \cos 3x}{9}$$

$$- (25/9) \int e^{5x} \cos 3x\, dx$$

It looks like we will be going forever. However notice the original integral and the last integral are the same except for a constant. Call the original integral I (for integral of course). The last line becomes . . .

$$I = \frac{e^{5x} \sin 3x}{3} + \frac{5e^{5x} \cos 3x}{9} - (25/9)I \qquad \text{now } I = (9/9)I$$

$$(34/9)I = \frac{e^{5x} \sin 3x}{3} + \frac{5e^{5x} \cos 3x}{9}$$

$$\text{so}\quad I = \int e^{5x} \cos 3x\, dx = (9/34)\left(\frac{e^{5x} \sin 3x}{3} + \frac{5e^{5x} \cos 3x}{9}\right) + C$$

$$= \frac{3e^{5x} \sin 3x}{34} + \frac{5e^{5x} \cos 3x}{34} + C$$

*Product of 3 minus signs is a minus.

Note: You do not have to multiply out the last step, but I did to show you that doing the problem 2 ways gives the same answer. Also note you do not have to do the problem 2 ways. I want to show you both ways give the same answer, and I am also a little crazy to do it both ways!!

$\int e^{5x} \cos 3x\, dx.$ $dv = e^{5x}.$ $v = e^{5x}/5.$ $u = \cos 3x.$ $du = -3 \sin 3x\, dx.$

$$\int \overset{u}{\cos 3x}\, \overset{dv}{e^{5x}}\, dx = \frac{\overset{u}{\cos 3x}\overset{v}{e^{5x}}}{5} - \int \overset{v}{\frac{e^{5x}}{5}}\, \overset{du}{(-3 \sin 3x)\, dx}$$

$$\int \overset{dv}{e^{5x}}\, \overset{u}{\cos 3x}\, dx = \frac{\cos 3xe^{5x}}{5} + (3/5) \int \overset{dV}{e^{5x}}\, \overset{U}{\sin 3x}\, dx$$

$dV = e^{5x}.$ $V = e^{5x}/5.$ $U = \sin 3x.$ $dU = 3 \cos 3x\, dx.$

$$\int e^{5x} \cos 3x\, dx = \frac{e^{5x} \cos 3x}{5} + \frac{\overset{V}{(3/5)e^{5x}}\, \overset{U}{\sin 3x}}{5} - (3/5) \int \frac{\overset{V}{e^{5x}}\, \overset{dU}{3 \cos 3x\, dx}}{5}$$

$$= \frac{e^{5x} \cos 3x}{5} + \frac{3e^{5x} \sin 3x}{25} - (9/25) \int e^{5x} \cos 3x\, dx$$

$$I = \frac{e^{5x} \cos 3x}{5} + \frac{3e^{5x} \sin 3x}{25} - (9/25)I \qquad I = (25/25)I$$

$$(34/25)I = \frac{e^{5x} \cos 3x}{5} + \frac{3e^{5x} \sin 3x}{25}$$

$$I = \int e^{5x} \cos 3x\, dx = (25/34)\left(\frac{e^{5x} \cos 3x}{5} + \frac{3e^{5x} \sin 3x}{25}\right) + C$$

$$= \frac{5e^{5x} \cos 3x}{34} + \frac{3e^{5x} \sin 3x}{34} + C$$

Our 2 answers check. Now that I've done it 2 ways to show you that both ways give the same answer, I will never, never do it twice again!!!!!!!!

The last integration by parts, unless I think of another, is the integral of $\sec^3 x$. I think it more properly belongs later (Example 10).

The next section involves integrals of trig functions. It is absolutely essential that you know the trig identities and integrals we listed before.

Let's consider integrals of the form $\sin^m x \cos^n x$.

EXAMPLE 5 m or n odd. $\int \sin^6 x \cos^3 x\, dx.$

The technique is to break off the trig function that is to an odd power (if both are, break off the 1 to the lower degree), and write all the others using the identity $\sin^2 x + \cos^2 x = 1$.

$$\int \sin^6 x \cos^3 x\, dx = \int (\sin^6 x)(1 - \sin^2 x)(\cos x)\, dx$$

$$u = \sin x;\ du = \cos x\, dx$$

$$= \int u^6(1 - u^2)\, du = \int (u^6 - u^8)\, du$$

$$= \frac{u^7}{7} - \frac{u^9}{9} + C$$

so $\displaystyle\int \sin^6 x \cos^3 x\, dx = \frac{\sin^7 x}{7} - \frac{\sin^9 x}{9} + C$

Pretty simple, eh? However, when m,n are to an even power, the integrals are usually much longer.

EXAMPLE 6 $\int \sin^6 x\, dx$. $m = 6$, $n = 0$ . . . both exponents even.

In doing these problems, we will use the half-angle formulas:

$\sin^2 x = (1 - \cos 2x)/2$ and $\cos^2 x = (1 + \cos 2x)/2$

$$\int \sin^6 x\, dx = \int (\sin^2 x)(\sin^2 x)(\sin^2 x)\, dx$$

$$= \int \left(\frac{1 - \cos 2x}{2}\right)\left(\frac{1 - \cos 2x}{2}\right)\left(\frac{1 - \cos 2x}{2}\right) dx$$

$$= \frac{1}{8}\int dx - \frac{3}{8}\int \cos 2x\, dx + \frac{3}{8}\int \cos^2 2x\, dx$$

$$- \frac{1}{8}\int \cos^3 2x\, dx$$

$$= \frac{1}{8}x - \frac{3 \sin 2x}{16} + A + B$$

$$\text{Integral A} = \frac{3}{8}\int \cos^2 2x\, dx = \frac{3}{8}\int \frac{1 + \cos 4x}{2}\, dx$$

$$= \frac{3}{16}\int 1\, dx + \frac{3}{16}\int \cos 4x\, dx$$

$$= \frac{3}{16}x + \frac{3 \sin 4x}{64}$$

$$\text{Integral B} = -\frac{1}{8}\int \cos^3 2x\, dx$$

$$= -\frac{1}{8}\int (1 - \sin^2 2x)(\cos 2x)\,dx$$

$u = \sin 2x$; $du = 2\cos 2x\,dx$

$$= -\frac{1}{16}\int (1 - \sin^2 2x)(2\cos 2x)\,dx$$

$$= -\frac{1}{16}\int (1 - u^2)\,du = -\frac{1}{16}\left(u - \frac{u^3}{3}\right)$$

$$= -\frac{1}{16}\left(\sin 2x - \frac{\sin^3 2x}{3}\right) = -\frac{\sin 2x}{16} + \frac{\sin^3 2x}{48}$$

Combining all parts of the integral, we get

$$\int \sin^6 x\,dx = \frac{x}{8} - \frac{3\sin 2x}{16} + \frac{3x}{16} + \frac{3\sin 4x}{64}$$

$$- \frac{\sin 2x}{16} + \frac{\sin^3 2x}{48} + C$$

$$= \frac{5x}{16} - \frac{\sin 2x}{4} + \frac{\sin^3 2x}{48} + \frac{3\sin 4x}{64} + C$$

Quite a problem. It certainly is much nicer if the sin or cos has an odd exponent.

We now examine the integrals involving $\tan^m x \sec^n x$. Before we start we will make 2 notes: whatever we say for tan-sec goes for cot-csc, and also notice that tan-sec and cot-csc are grouped together whether for trig identities or trig integrals.

EXAMPLE 7 $\int \tan^5 x \sec^5 x\,dx$

m,n odd. $u = \sec x$. $du = \tan x \sec x\,dx$. Break off 1 tan and 1 sec. Write each remaining $\tan^2 x$ as $\sec^2 x - 1$.

$$\int \tan^5 x \sec^5 x\,dx = \int (\sec^2 x - 1)(\sec^2 x - 1)(\sec^4 x)(\tan x \sec x)\,dx$$

$$= \int (u^2 - 1)(u^2 - 1)(u^4)\,du$$

$$= \int (u^8 - 2u^6 + u^4)\,du = \frac{u^9}{9} - \frac{2u^7}{7} + \frac{u^5}{5} + C$$

$$= \frac{\sec^9 x}{9} - \frac{2\sec^7 x}{7} + \frac{\sec^5 x}{5} + C$$

EXAMPLE 8 $\int \sec^4 x\,dx$

m,n even. This is one of my favorites. If m,n are even and in this problem even though there is no tangent, we let $u = \tan x$, $du = \sec^2 x\,dx$, break off 2 of the secants, and write the remaining $\sec^2 x$ as $\tan^2 x + 1$.

$$\int \sec^4 x\,dx = \int (1 + \tan^2 x)(\sec^2 x)\,dx \qquad u = \tan x;\ du = \sec^2 x\,dx$$

$$= \int (1 + u^2)\,du = u + u^3/3 + C$$

$$= \tan x + (\tan^3 x)/3 + C$$

EXAMPLE 9 $\int \tan^3 x \sec^4 x\,dx$

m (power of tan) is odd and n (power of sec) is even. We can let $u = \tan x$ or $v = \sec x$. We will do it both ways. Neither is too bad.

EXAMPLE 9A $\int \tan^3 x \sec^4 x\,dx \qquad u = \tan x,\ du = \sec^2 x\,dx$

$$= \int (\tan^3 x)(\tan^2 x + 1)\sec^2 x\,dx$$

$$= \int u^3(u^2 + 1)\,du = \int (u^5 + u^3)\,du = \frac{u^6}{6} + \frac{u^4}{4} + C$$

$$= \frac{\tan^6 x}{6} + \frac{\tan^4 x}{4} + C$$

EXAMPLE 9B $\int \tan^3 x \sec^4 x\,dx \qquad v = \sec x;\ dv = \tan x \sec x\,dx$

$$= \int (\sec^2 x - 1)(\sec^3 x)(\tan x \sec x)\,dx$$

$$= \int (v^2 - 1)(v^3)\,dv = \int (v^5 - v^3)\,dv = \frac{v^6}{6} - \frac{v^4}{4} + C$$

$$= \frac{\sec^6 x}{6} - \frac{\sec^4 x}{4} + C$$

You might try to show that the answers to 9A and 9B are the same using the identity $\sec^2 x = \tan^2 x + 1$.

EXAMPLE 10 $\int \sec^3 x\,dx$

This is the worst case: the power of $\tan x$, m, is even—$m = 0$—and the power of the $\sec x$, n, is odd—$n = 3$. All cases where m is even and n is odd are done by integrating by parts. They get long fast as the powers of m and n increase and all involve the same tricks.

$$\int \sec^3 x\,dx = \int \overset{u}{\sec x}\ \overset{dv}{\sec^2 x\,dx} = \overset{u}{\sec x}\ \overset{v}{\tan x} - \int \overset{v}{(\tan x)}\ \overset{du}{\tan x \sec x\,dx}$$

$$= \sec x \tan x - \int [(\sec^2 x - 1)\sec x]\,dx$$

$$= \sec x \tan x - \int \sec^3 x\,dx + \int \sec x\,dx$$

$$= \sec x \tan x - \int \sec^3 x\,dx + \ln|\sec x + \tan x|$$

Solving for the unknown integral, we get

$$\int \sec^3 x\,dx = \frac{\sec x \tan x + \ln|\sec x + \tan x|}{2} + C$$

EXAMPLE 11 $\int \tan^2 x \csc x\,dx$

Anytime you have a "mixed" integral, that is, where tan is not with the sec, you will have to use trig identities and usually tricks and sometimes long problems involving techniques that may not have been done here yet. This one I've given is a rather mild one.

$$\int \tan^2 x \csc x\,dx = \int \frac{\sin^2 x}{\cos^2 x}\,\frac{1}{\sin x}\,dx \qquad u = \cos x;\ du = -\sin x\,dx$$

$$= \int \frac{-du}{u^2} = \int -u^{-2}\,du$$

$$= 1/u = 1/\cos x + C = \sec x + C$$

We now have integrals involving square roots. Our goal is to get rid of the radicals. The first area here is trig substitutions.

Type I: $(a^2 - x^2)^{1/2}$. We use $x = a \sin u$ ($dx = a \cos u\,du$).
$(a^2 - x^2)^{1/2} = (a^2 - a^2 \sin^2 u)^{1/2} = [a^2(1 - \sin^2 u)]^{1/2} = a \cos u$

and the square root is gone. Here are the other two cases.

Type II: $(a^2 + x^2)^{1/2}$. We use $x = a \tan u$ ($dx = a \sec^2 u\,du$).
$(a^2 + x^2)^{1/2} = (a^2 + a^2 \tan^2 u)^{1/2} = [a^2(1 + \tan^2 a)]^{1/2} = a \sec u$.

Type III: $(x^2 - a^2)^{1/2}$. We use $x = a \sec u$ ($dx = a \tan u \sec u\,du$).
$(x^2 - a^2)^{1/2} = (a^2 \sec^2 u - a^2) = [a^2(\sec^2 u - 1)]^{1/2} = a \tan u$.

I have demonstrated each of the three types. However it is essential that you know by sight what the answer is without substituting. Otherwise the problems will take forever. If I wake you up in the middle of the night and ask you, "What do you get if you have $(7-x^2)^{1/2}$?" You should instantly say, "Square root of 7 cosine u—now let me go back to sleep!!!!"

EXAMPLE 12 $\int \frac{dx}{x^2(16-x^2)^{1/2}}$ $\quad x = 4 \sin u;\ dx = 4 \cos u\, du$

$$= \int \frac{4 \cos u\, du}{(16 \sin^2 u)(4 \cos u)}$$

$$= \int \frac{du}{16 \sin^2 u} = \int \frac{\csc^2 u\, du}{16} = \frac{-\cot u}{16} + C$$

We didn't start with u; we started with x. We must draw a triangle with $x = 4 \sin u$. $\sin u = x/4$. Note the missing side will be what the square root is.

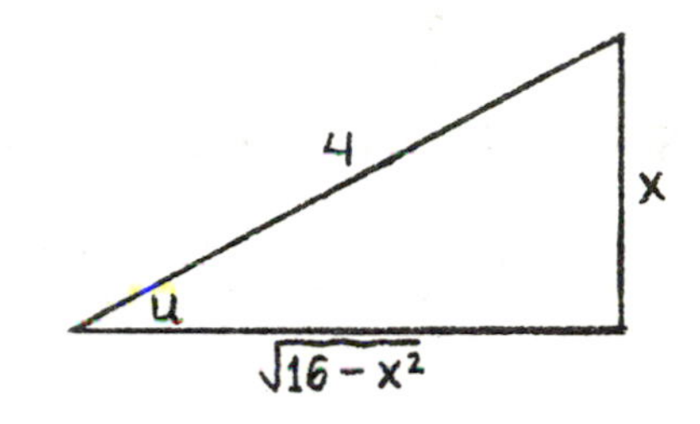

$$\frac{-\cot u}{16} + C = -\frac{1}{16}\left(\frac{\sqrt{16-x^2}}{x}\right) + C$$

EXAMPLE 13 $\int \frac{x^2\, dx}{(16-x^2)^{1/2}}$

This problem appears to be exactly the same as the last except the x^2 is on top instead of the bottom. This problem is given to show that the techniques are different even in problems that look the same—some longer, some shorter, some easier, some harder. The kind of problem is known only after lots of study. Do them and hope they are short and easy.

Again let $x = 4 \sin u$.
$dx = 4 \cos u\, du$

$$\int \frac{x^2\, dx}{(16-x^2)^{1/2}}$$

$$= \int \frac{(16 \sin^2 u)(4 \cos u)\, du}{4 \cos u} = \int 16 \sin^2 u\, du = \int \frac{16(1-\cos 2u)\, du}{2}$$

$$= \int 8\, du - \int 8 \cos 2u\, du$$

$$= 8u - 4 \sin 2u$$

$$= 8u - 8 \sin u \cos u$$

$$= 8 \arcsin (x/4) - 8\left[\frac{x}{4}\ \frac{(16-x^2)^{1/2}}{4}\right] + C$$

Notes:

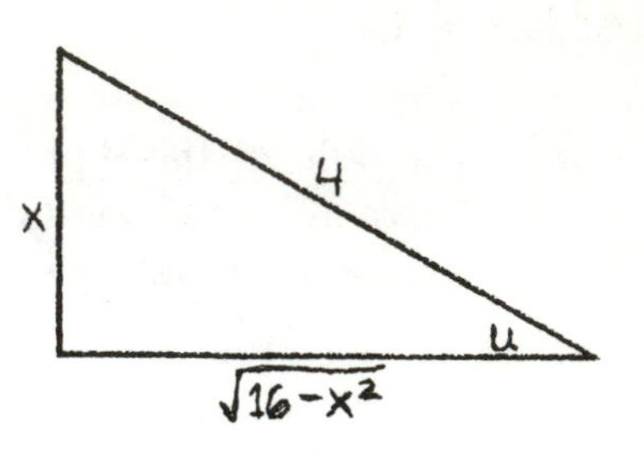

$x = 4 \sin u, \quad x/4 = \sin u, \quad \sin^{-1}(x/4) = u$

$-4 \sin 2u = -4(2 \sin u \cos u) = -8 \sin u \cos u$

$\frac{x}{4} = \sin u \quad \text{so} \quad \frac{\sqrt{16 - x^2}}{4} = \cos u$

As you can see, these two problems are quite different although looking basically the same.

We will finish by showing that the area of a circle really is pi r squared. We will find one-quarter the area of the circle $x^2 + y^2 = r^2$, and then multiply it by 4.

EXAMPLE 14

$$4\int_0^r (r^2 - x^2)^{1/2}\,dx \qquad \mathbf{x = r \sin u.\ dx = r \cos u\ du}$$

x = r, r = r sin u, 1 = sin u, u = π/2; x = 0, 0 = r sin u, 0 = sin u, u = 0

$$= 4\int_0^{\pi/2} (r \cos u)(r \cos u)\,du$$

$$= 4r^2\int_0^{\pi/2} \cos^2 u\,du$$

$$= 4r^2\int_0^{\pi/2} \frac{(1 + \cos 2u)\,du}{2}$$

$$= 2r^2\int_0^{\pi/2} (1 + \cos 2u)\,du$$

$$= 2r^2(u + \sin 2u)\Big[_0^{\pi/2} = 2r^2\left[\left(\frac{\pi}{2} + \sin \pi\right) - (0 - \sin 0)\right]$$

$$= 2r^2(\pi/2) = \pi r^2!!$$

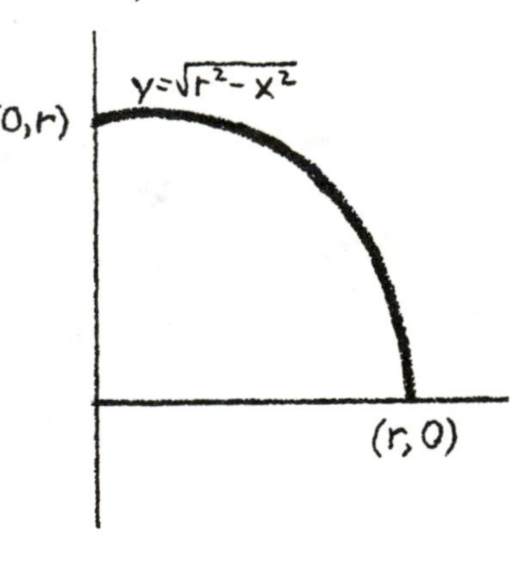

The area of a circle really is pi r squared, and you haven't been lied to all these years. It's nice to know.

EXAMPLE 15

$$\int \frac{dx}{(25 + x^2)^{1/2}} \qquad \mathbf{x = 5 \tan u.\ dx = 5 \sec^2 u\ du}$$

$$= \int \frac{5 \sec^2 u\,du}{5 \sec u}$$

$$= \int \sec u\,du$$

$$= \ln|\sec u + \tan u| + C$$

$$= \ln\left|\frac{\sqrt{25 + x^2}}{5} + \frac{x}{5}\right| + C$$

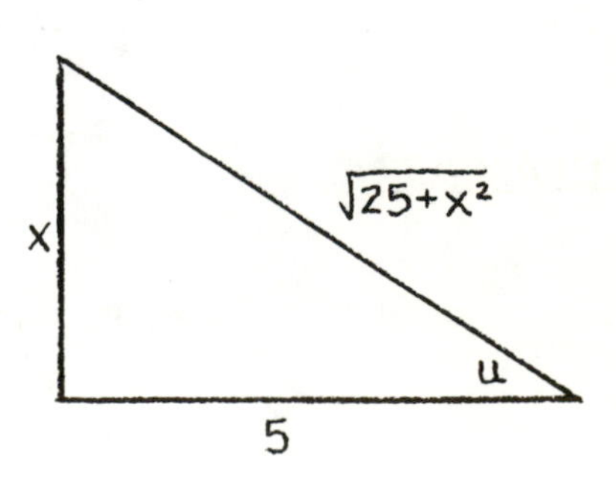

EXAMPLE 16

$$\int \frac{dx}{x^2(x^2-7)^{1/2}}$$

$x = 7^{1/2}\sec u$. $dx = 7^{1/2}\tan u \sec u\, du$

$$= \int \frac{7^{1/2}\tan u \sec u\, du}{(7\sec^2 u)(7^{1/2}\tan u)}$$

$$= \int \frac{du}{7\sec u} = \int \frac{\cos u\, du}{7}$$

$$= \frac{1}{7}\sin u = \frac{(x^2-7)^{1/2}}{7x} + C$$

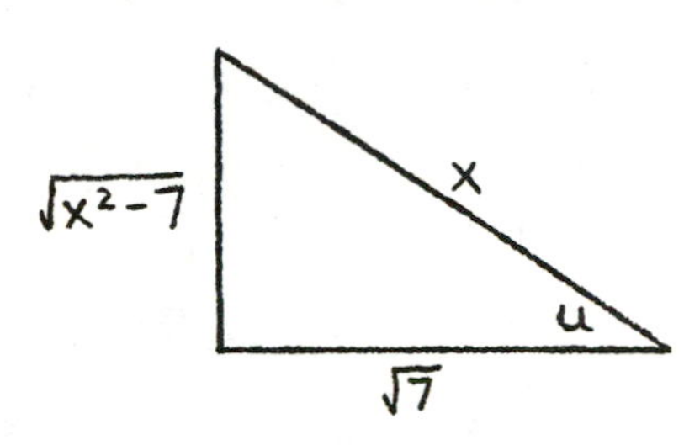

EXAMPLE 17

$$\int \frac{dx}{(x^2+9)^2}$$

Sometimes you use a trig substitution even though there is no square root. $x = 3\tan u$. $dx = 3\sec^2 u\, du$

$$= \int \frac{3\sec^2 u\, du}{81\sec^4 u} = \int \frac{\cos^2 u\, du}{27} = \int \frac{(1+\cos 2u)\, du}{54}$$

$$= (1/54)u + (\sin 2u)/108$$

$$= (1/54)u + (\sin u \cos u)/54 + C$$

$$= \frac{1}{54}\tan^{-1}\frac{x}{3} + \frac{1}{54}\left(\frac{x}{\sqrt{x^2+9}}\cdot\frac{3}{\sqrt{x^2+9}}\right) + C$$

$$= \frac{1}{54}\tan^{-1}\frac{x}{3} + \frac{x}{18(x^2+9)} + C$$

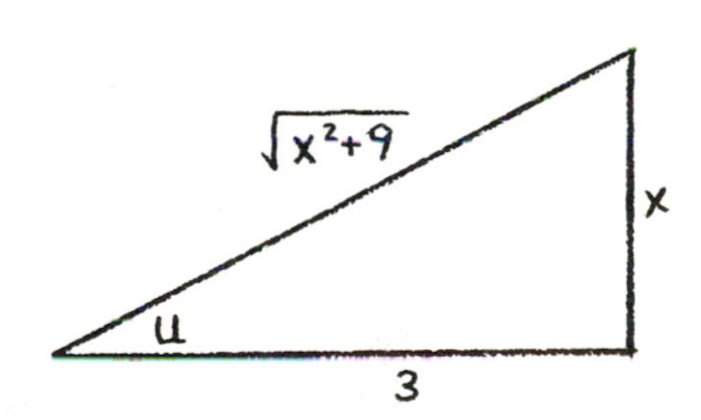

One last example. Warning!!! A trig substitution may work, but another method may be a lot easier.

EXAMPLE 18

$$\int \frac{2x\, dx}{(x^2+9)^{100}}$$

$x = 3\tan u$ will work. But you may not finish for 2 years. The best? $u = x^2 + 9$. $du = 2x\, dx$

$$= \int u^{-100}\, du = \frac{u^{-99}}{-99} + C = \frac{-1}{99(x^2+9)^{99}} + C$$

Another group of integrals are related to the last group. They can be very involved, but we will do 2 moderate ones.

EXAMPLE 19

$$\int \frac{(2x-3)\, dx}{3x^2 - 18x + 75}$$

We do this by completing the square. $3x^2 - 18x + 75 = 3(x^2 - 6x + 25) = 3(x^2 - 6x + 9 + 16) = 3[(x-3)^2 + 16]$

$$= \int \frac{(2x-3)\, dx}{3[(x-3)^2+16]}$$

$u = x - 3$. $x = u + 3$. $2x - 3 = 2(u+3) - 3 = 2u + 3$. $du = dx$

$$= \frac{1}{3}\int \frac{(2u+3)\,du}{u^2+16}$$

Now split the integral.

$$= \frac{1}{3}\int \frac{2u\,du}{u^2+16} + \int \frac{1\,du}{u^2+16}$$

Both of these integrals should be known by sight!

$$= \frac{1}{3}\ln(u^2+16) + \tfrac{1}{4}\tan^{-1}(u/4) + C$$

$$= \frac{1}{3}\ln(x^2-6x+25) + \tfrac{1}{4}\tan^{-1}[(x-3)/4] + C$$

since $u = x - 3$ and $u^2 + 16 = (x-3)^2 + 16 = x^2 - 6x + 9 + 16 = x^2 - 6x + 25$.

EXAMPLE 20

$$\int \frac{dx}{(15+2x-x^2)^{1/2}}$$

Again we complete the square. $15 + 2x - x^2 = -1(x^2 - 2x) + 15 = -1(x^2 - 2x + 1) + 16 = 16 - (x-1)^2$

$$= \int \frac{dx}{[16-(x-1)^2]^{1/2}}$$

Again you should be able to tell this is an arc sin, that is, $\int dx/(a^2 - u^2)^{1/2}$ where $a = 4$ and $u = x - 1$.

$$= \text{arc sin}\,[(x-1)/4] + C$$

We now do the section I like the least. It is uninteresting, unimaginative, frequently overly long, and . . . necessary. Unless we have only linear factors, it is best to avoid this technique if possible.

We wish to do the integrals by partial fractions. Suppose we have $R(x) = P(x)/Q(x)$, where the degree of $P(x)$ is less than the degree of $Q(x)$. We wish to break up $R(x)$ into simpler rational fractions; each piece is called a partial fraction. There will be 1 or more pieces for each linear factor $x + a$ or quadratic factor $x^2 + b^2$ of $Q(x)$. Here's how it looks in a particular case:

$$\frac{P(x)}{x^2(x-3)^3(x^2+4)^2} = \underbrace{\frac{A}{x} + \frac{B}{x^2}}_{\text{2 from } x} + \underbrace{\frac{C}{x-3} + \frac{D}{(x-3)^2} + \frac{E}{(x-3)^3}}_{\text{3 from } (x-3)}$$

$$+ \underbrace{\frac{Fx+G}{x^2+4} + \frac{Hx+I}{(x^2+4)^2}}_{\text{2 from } (x^2+4)}$$

Notice that each linear factor gives pieces with constants on top, and each quadratic factor gives pieces with first-degree polynomials on top. The bottoms of the partial fractions are powers of the factors running from 1 to the power that occurs in $Q(x)$. The constants A,B,C,D,E,F,G,H, and I have to be solved for, which I hope you never have to do. If you added all the fractions on the right, you would get the left fraction.

One more thing: Suppose $Ax^3 + Bx^2 + Cx + D = 4x^3 - 7x - 1$. Two polynomials are equal if their coefficients match. So $A = 4$, $B = 0$, $C = -7$, $D = -1$.

There are a number of techniques that will allow you to solve for A,B,C, etc. Two of them (combinations of) will serve us best.

EXAMPLE 21

$$\int \frac{x^3 - 7x + 18}{x^2 - 9}\,dx$$

Since the degree of top ≥ degree of bottom, long divide the bottom into the top until degree of top less than degree of bottom

$$= \int \left(x + \frac{2x + 18}{x^2 - 9}\right) dx$$

Look at fractional part only

$$\frac{2x + 18}{x^2 - 9} = \frac{A}{x - 3} + \frac{B}{x + 3}$$

We will solve for A and B in 2 different ways. We now add the fractions and equate the tops since the bottoms are the same

$$2x + 18 = A(x + 3) + B(x - 3)$$

Method 1

$$A(x + 3) + B(x - 3) = 2x + 18$$

Multiply out left side and group terms

$$(A + B)x + (3A - 3B) = 2x + 18$$

Now match coefficients

$$A + B = 2$$
$$3A - 3B = 18$$

Solve 2 equations in 2 unknowns. It is really important for your algebra to be good

$$3A + 3B = 6$$
$$3A - 3B = 18$$
$$6A = 24$$
$$A = 4$$

Substitute in either equation

$$B = -2$$

Method 2

$A(x+3)+B(x-3)=2x+18$

This is true for all values of x. If we substitute x = 3 in both sides and then x = −3 in both sides, we will get both A and B with no work

If $x=3$, $A(3+3)+B(3-3)=2(3)+18$. $6A=24$. $A=4$.

If $x=-3$, $A(-3+3)+B(-3-3)=2(-3)+18$. $-6B=12$. $B=-2$.

This way is so much easier—why don't we always use it? It is only perfect if we have all linear factors to the first power. Otherwise it will not totally work. If there are no linear factors, you can't use this method. That is why both methods are needed. Let us finally finish the problem!

$$\int \frac{x^3-7x+18\,dx}{x^2-9} = \int x\,dx + \int \frac{4\,dx}{x-3} + \int \frac{-2\,dx}{x+3}$$

$$= \frac{x^2}{2} + 4\ln|x-3| - 2\ln|x+3| + C$$

Note how easy the calculus part is. The algebra can be overwhelming.

EXAMPLE 22 $\displaystyle\int \frac{9x^2-5x+19\,dx}{(x^2+5)(x-2)}$

Notice the degree of the top (2) is less than the degree of the bottom (3), so long division is not necessary. The bottom is already factored. There is 1 quadratic factor and 1 linear factor. The form is

$$\frac{Ax+B}{x^2+5} + \frac{C}{x-2}$$

$$\frac{9x^2-5x+19}{(x^2+5)(x-2)} = \frac{Ax+B}{x^2+5} + \frac{C}{x-2} = \frac{(Ax+B)(x-2)+C(x^2+5)}{(x^2+5)(x-2)}$$

We now multiply out the top on the right, and set the left numerator equal to the right numerator.

$$9x^2-5x+19 = (A+C)x^2 + (-2A+B)x + (-2B+5C)$$

(1) $A+C=9$

(2) $-2A+B=-5$

(3) $-2B+5C=19$

These 3 equations in 3 unknowns are not bad but not particularly nice to solve. So we can use the other technique. Going back to the original top on the right, we have $(Ax + B)(x - 2) + C(x^2 + 5) = 9x^2 - 5x + 19$. There is only 1 linear factor, $x - 2$, but it is enough. Substituting $x = 2$ in this equation we get $(A(2) + B)(2 - 2) + C(2^2 + 5) = 9(2)^2 - 5(2) + 19$. From this we get $9C = 45$ or $C = 5$. Substituting $C = 5$ into (1), we get $A = 4$. Substituting $A = 4$ into (2), we get $B = 3$.

$$\int \frac{9x^2 - 5x + 19}{(x^2 + 5)(x - 2)}\,dx = \int \left(\frac{4x + 3}{x^2 + 5} + \frac{5}{x - 2}\right) dx$$

Splitting the first fraction on the right,

$$\int \frac{9x^2 - 5x + 19}{(x^2 + 5)(x - 2)}\,dx = 2\int \frac{2x\,dx}{x^2 + 5} + \int \frac{3\,dx}{x^2 + 5} + \int \frac{5\,dx}{x - 2}$$

$$= 2 \ln (x^2 + 5) + (3/5^{1/2}) \arctan (x/5^{1/2}) + 5 \ln |x - 2| + C$$

EXAMPLE 23

$$\int \frac{(6x^2 + 17x - 5)\,dx}{(x - 3)(x + 2)^2}$$

We have 2 linear factors, and 1 is to the second power—soooo . . . . .

$$\frac{6x^2 + 17x - 5}{(x - 3)(x + 2)^2} = \frac{A}{x - 3} + \frac{B}{x + 2} + \frac{C}{(x + 2)^2}$$

$$= \frac{A(x + 2)^2 + B(x + 2)(x - 3) + C(x - 3)}{(x - 3)(x + 2)^2}$$

(1) $6x^2 + 17x - 5 = A(x + 2)^2 + B(x + 2)(x - 3) + C(x - 3)$

(2) $= A(x^2 + 4x + 4) + B(x^2 - x - 6) + x(C - 3)$ **Multiply and group; we get**

(3) $A + B = 6$ (4) $4A - B + C = 17$ (5) $4A - 6B - 3C = -5$

Now there are two good numbers, -2 and 3, but as we will see, 3 is enough. Putting $x = 3$ into both sides of (1), we get $A = 4$. Putting $A = 4$ into (3), we get $B = 2$. Putting $B = 2$ and $A = 4$ into (4) or (5), we get $C = 3$.

$$\int \frac{6x^2 + 17x - 5\,dx}{(x + 2)^2(x - 3)} = \int \frac{4\,dx}{x - 3} + \int \frac{2\,dx}{x + 2} + \int \frac{3\,dx}{(x + 2)^2}$$

$$= 4 \ln |x - 3| + 2 \ln |x + 2| - 3/(x + 2) + C$$

The last part of this long integral chapter is called—

Miscellaneous

"Miscellaneous" means that anything that doesn't fit into any other part goes into this part and makes it more miserable for you.

EXAMPLE 24 $\int x(x+1)^{80}\,dx$

Sometimes the simplest substitutions work. We let $u = x + 1$. $du = dx$. This transfers the power to the monomial and allows us to multiply out the expression $(x = u - 1)$.

$$\int x(x+1)^{80}\,dx = \int (u-1)u^{80}\,du = \int (u^{80} - u^{81})\,du$$

$$= (u^{81})/81 - (u^{82})/82 + C = \frac{(x+1)^{81}}{81} - \frac{(x+1)^{82}}{82} + C$$

You sharp-eyed readers will note that there are at least 2 other ways to do this problem. The first is to multiply out the original. This is dismissed on grounds of sanity. The second is by parts.

EXAMPLE 25 $\int \frac{dx}{x^{1/2} - x^{1/3}}$

This one is perhaps the easiest to identify. LCD for $\frac{1}{2}$, $\frac{1}{3}$ is 6. So . . . we let $u = x^{1/6}$. $u^2 = (x^{1/6})^2 = x^{1/3}$. $u^3 = (x^{1/6})^3 = x^{1/2}$. Finally $u^6 = x$; so $6u^5\,du = dx$. Substituting we get . . .

$$\int \frac{dx}{x^{1/2} - x^{1/3}} = \int \frac{6u^5\,du}{u^3 - u^2} = 6\int \frac{u^5\,du}{u^2(u-1)}$$

$$= 6\int \frac{u^3\,du}{u-1}$$

$$= 6\int [u^2 + u + 1 + 1/(u-1)]\,du$$

$$= 6(u^3/3 + u^2/2 + u + \ln|u-1|) + C$$

$$= 2(x^{1/6})^3 + 3(x^{1/6})^2 + 6x^{1/6} + \ln|x^{1/6} - 1| + C$$

$$= 2x^{1/2} + 3x^{1/3} + 6x^{1/6} + \ln|x^{1/6} - 1| + C$$

$$\begin{array}{r l}
 & \underline{u^2 + u + 1} + 1/(u-1) \\
u - 1\,) & u^3 \\
 & \underline{u^3 - u^2} \\
 & u^2 \\
 & \underline{u^2 - u} \\
 & u \\
 & \underline{u - 1} \\
 & 1
\end{array}$$

Note the nice pattern of the coefficients and the exponents of this answer. Well, *I* like it!

EXAMPLE 26 $\int \frac{x^3\,dx}{(x^2+4)^{1/2}}$

This integral can be done in 2 new ways, both of which are useful.

EXAMPLE 26, method A Let u = the whole radical.

$u=(x^2+4)^{1/2}$. $u^2=x^2+4$. $x^2=u^2-4$. $2x\,dx=2u\,du$ or $x\,dx=u\,du$. Substituting,

$$\int \frac{x^3\,dx}{(x^2+4)^{1/2}} = \int \frac{x^2x\,dx}{(x^2+4)^{1/2}} = \int \frac{(u^2-4)u\,du}{u}$$

$$= \int (u^2-4)\,du = u^3/3 - 4u + C$$

$$= \frac{(x^2+4)^{3/2}}{3} - 4(x^2+4)^{1/2} + C$$

EXAMPLE 26, method B Let v = what's under the radical sign.

$v=x^2+4$. $x^2=v-4$. $2x\,dx=dv$. Substituting,

$$\int \frac{x^3\,dx}{(x^2+4)^{1/2}} = \frac{1}{2}\int \frac{x^2 2x\,dx}{(x^2+4)^{1/2}} = \frac{1}{2}\int \frac{(v-4)\,dv}{v^{1/2}}$$

$$= \frac{1}{2}\int (v^{1/2}-4v^{-1/2})\,dv = \frac{1}{2}\,[(2/3)v^{3/2} - 8v^{1/2}] + C$$

$$= (1/3)(x^2+4)^{3/2} - 4(x^2+4)^{1/2} + C$$

The sharp-eyed reader will discover there are many, many other ways to do this problem. When a publisher becomes smart enough to publish this book, this problem will become a contest.

Finally the weirdest miscellaneous of all! If we have sin x or cos x in the denominator, and if they are added or subtracted to each other (with one being multiplied by a number or being added or subtracted by itself from a number), then the substitution is $u = \tan(x/2)$!!!!!!!! I don't know who discovered it or why, but it works.

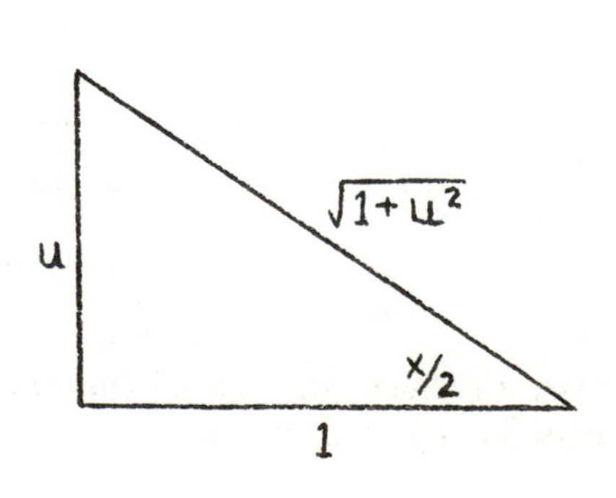

Let us derive all the parts. Otherwise you would never believe it. Let $\tan(x/2) = u = u/1$. Draw the triangle for x/2. We get $\sin(x/2) = u/(1+u^2)^{1/2}$ and $\cos(x/2) = 1/(1+u^2)^{1/2}$.

$$\sin x = 2 \sin (x/2) \cos (x/2)$$

$$= 2 \frac{u}{(1+u^2)^{1/2}} \frac{1}{(1+u^2)^{1/2}} = \frac{2u}{1+u^2}$$

Draw triangle for angle x.

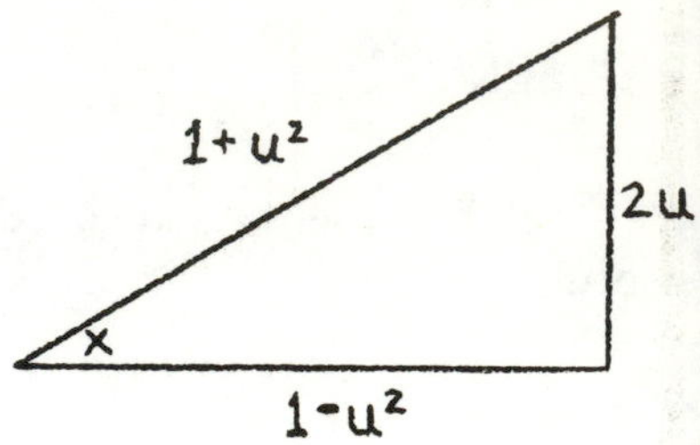

$$\cos x = \frac{1-u^2}{1+u^2}$$

Finally if $u = \tan (x/2)$, then $\tan^{-1} u = x/2$. Taking differentials, we get

$$\frac{2\,du}{1+u^2} = dx$$

In summary,

$$u = \tan (x/2) \qquad \frac{2\,du}{(1+u^2)} = dx \qquad \sin x = \frac{2u}{1+u^2} \qquad \cos x = \frac{1-u^2}{1+u^2}$$

EXAMPLE 27 $\displaystyle\int \frac{1+\sin x}{1+\cos x}\,dx$

$$= \int \frac{1+\dfrac{2u}{1+u^2}}{1+\dfrac{1-u^2}{1+u^2}} \frac{2\,du}{1+u^2}$$

$$= \int \frac{\dfrac{u^2+2u+1}{1+u^2}}{\dfrac{1+u^2+1-u^2}{1+u^2}} \frac{2\,du}{1+u^2}$$

$$= \int \frac{u^2+2u+1}{2} \frac{2\,du}{1+u^2}$$

$$= \int \frac{u^2+2u+1}{u^2+1}\,du$$

Trick—split the integral instead of long division

$$= \int \frac{(u^2+1)\,du}{u^2+1} + \int \frac{2u\,du}{1+u^2} = \int 1\,du + \int \frac{2u\,du}{1+u^2}$$

$$= u + \ln (u^2+1) + C = \tan (x/2) + \ln [\tan^2 (x/2) + 1] + C$$

or, if you want to be fancy, $= \tan (x/2) + \ln [\sec^2 (x/2)] + C$.

7. L'HOPITAL'S RULE

There was a little on L'Hopital's rule in the volume before. Let us give a complete discussion since we now know logs.

L'Hopital's rule: 1. $\lim_{x\to a} \frac{f(x)}{g(x)} = 0/0$ or ∞/∞

2. $\lim_{x\to a} \frac{f'(x)}{g'(x)} = L$

then $\lim_{x\to a} \frac{f(x)}{g(x)} = L$

a is any number or plus or minus infinity, and L could be any number or infinity.

This rule states if the original limit is zero over zero or infinity over infinity or can be made into that form, then by taking the derivative of the top and the derivative of the bottom, not the quotient rule, we can find the limit of the original.

EXAMPLE 1 $\lim_{x\to 0} \frac{x}{x^2+1} = \frac{0}{1} = 0$

No L'Hopital's rule.

EXAMPLE 2 $\lim_{x\to \infty} \frac{x^2+1}{1/x} = \frac{\infty}{0}$

which means the limit does not exist. Also no L'Hopital's rule.

EXAMPLE 3 $\lim_{x\to 0} \frac{\sin x}{x} = \frac{0}{0}$

Yay! L'Hopital's rule can be used.

$$\lim_{x\to 0} \frac{(\sin x)'}{(x)'} = \lim_{x\to 0} \frac{\cos x}{1} = 1/1 = 1 \qquad \text{so} \qquad \lim_{x\to 0} \frac{\sin x}{x} = 1$$

EXAMPLE 4 Also 0/0.

$$\lim_{x\to 1} \frac{\ln x}{x-1} = \lim_{x\to 1} \frac{(\ln x)'}{(x-1)'} = \lim_{x\to 1} \frac{(1/x)}{1} = 1 \qquad \text{so} \qquad \lim_{x\to 1} \frac{\ln x}{x-1} = 1$$

EXAMPLE 5 ∞/∞

Here we need to use L'Hopital's rule twice.

$$\lim_{x\to\infty} \frac{2x^2 - 3x + 1}{5 - 7x^2} = \lim_{x\to\infty} \frac{4x-3}{-14x} = \frac{4}{-14} = -2/7$$

$$\text{so} \qquad \lim_{x\to\infty} \frac{2x^2 - 3x + 1}{5 - 7x^2} = -2/7$$

EXAMPLE 6 $\lim_{x\to 0} (x \ln x)$

This turns out to be zero times infinity (minus infinity). We must rearrange it so it is either 0/0 or ∞/∞. We use a little trick to get it ∞/∞

$$\lim_{x\to 0} \frac{\ln x}{1/x} = \lim_{x\to 0} \frac{(\ln x)'}{(1/x)'} = \lim_{x\to 0} \frac{1/x}{-1/x^2} = \lim_{x\to 0} (-x) = 0$$

$$\text{so} \qquad \lim_{x\to 0} (x \ln x) = 0$$

EXAMPLE 7 $0 \times \infty.\ \lim_{x\to 0} (x \cot x) = \lim_{x\to 0} (x/\tan x)$ **Trig identity. Now it's 0/0**

$$\lim_{x\to 0} (x)'/(\tan x)' = 1/\sec^2 x = 1/1 = 1 \qquad \lim_{x\to 0} (x \cot x) = 1$$

EXAMPLE 8 $\lim_{x\to 0^+} (1/\sin x - 1/x)$ **Trick—add the fractions, get 0/0**

$$\lim_{x\to 0^+} \frac{x - \sin x}{x \sin x} = \lim_{x\to 0^+} \frac{(1 - \cos x)}{x \cos x + \sin x} = \lim_{x\to 0^+} \frac{\sin x}{2\cos x - (\sin x)x} = 0/2 = 0$$

Using L'Hopital's rule twice we get the original limit to be 0.

If you exclude the log examples, all of the preceding could have been done in the first semester. However the following examples require logs. Those requiring logs are of the form 0^0, ∞^0, and 1^∞.

EXAMPLE 9 0^0. $\lim_{x\to 0^+} x^{4x}$. $y = x^{4x}$. $\ln y = 4x \ln x \to 0$ by Example 6. So $y \to e^0 = 1$.

EXAMPLE 10 ∞^0. $\lim_{x\to\infty} x^{1/x}$. $y = x^{1/x}$. $\ln y = (1/x)\ln x$. ∞/∞.

$$\lim_{x\to\infty} \frac{(\ln x)'}{x'} = \lim_{x\to\infty} \frac{(1/x)}{1} = 0/1 = 0$$

Since $\ln y \to 0$, $y \to e^0 = 1$.

EXAMPLE 11 1^∞. $\lim_{x\to\infty} [1 + (1/x)]^x$.

$$y = \left(x + \frac{1}{x}\right)^x \quad \ln y = x \ln [1 + (1/x)] = \frac{\ln [1 + (1/x)]}{1/x} = 0/0$$

Taking derivatives top and bottom, we get

$$\frac{\dfrac{-1/x^2}{1 + (1/x)}}{-1/x^2} = \frac{1}{1 + (1/x)} \to 1$$

Thus $\ln y \to 1$ and so $y \to e^1 = e$.

Note: $\lim_{x\to\infty} (1 + [(a/bx)]^{cx} = e^{ac/b}$

EXAMPLE 12 0^∞ not L'Hopital's rule, because $0^k = 0$ for all positive k.

Note: We use L'Hopital's rule if we have $\frac{0}{0}$, $\frac{\infty}{\infty}$, $0 \cdot \infty$, $\infty - \infty$, 0^0, ∞^0, 1^∞. It's a no-no for $\frac{0}{\infty}$, $\frac{\infty}{0}$, $\infty \cdot \infty$, 0^∞, ∞^∞.

8. IMPROPER INTEGRALS

In discussing an improper integral, it would seem to be a good idea to recall what a "proper" integral is. In Calc I we defined the integral of f(x) from a to b this way: Break up the interval (a,b) into n parts. Let w_i be any point in the interval Δx_i. Form the sum $f(w_1)\Delta x_1 + f(w_2)\Delta x_2 + f(w_3)\Delta x_3 + \cdots + f(w_n)\Delta x_n$. Form the sum $\sum_{i=1}^{n} f(w_i)\Delta x_i$. If the limit exists as n goes to infinity and all the deltas go to zero, we have

$$\lim_{\substack{n\to\infty \\ \text{all } \Delta x\text{'s}\to 0}} \sum_{i=1}^{n} f(w_i)\Delta x_i = \int_a^b f(x)\,dx$$

We usually, at the start, take f(x) to be continuous, although that can be weakened. However, implied in the definition is that everything is *finite*, that is, both a and b are finite and f(x) is always finite. What happens if we have an infinity? In effect we close our eyes and pretend the infinity is not there. We then take the limit as we go to that infinity. If the limit gives us a single finite number, we will say the integral CONVERGES to that number. Otherwise the integral DIVERGES. Let us be more formal.

EXAMPLE 1 $\displaystyle\int_{-1}^{\infty} \frac{1\,dx}{1+x^2}$

We rewrite this as

$$\lim_{a\to\infty} \int_{-1}^{a} \frac{1\,dx}{1+x^2} = \lim_{a\to\infty} \tan^{-1} x \Big[_{-1}^{a} = \lim_{a\to a} [\tan^{-1} a - \tan^{-1}(-1)]$$

$$= \pi/2 - (-\pi/4) = \frac{3\pi}{4}$$

You might ask, "Are they all this easy?" In most books, the vast majority of the improper integrals are relatively easy, in order to make sure that you understand what an improper integral is without worrying about a complicated integral.

In summary, this integral CONVERGES to the value $3\pi/4$.

EXAMPLE 2 $\int_4^{\infty} 1/x^{1/2}\, dx$

We write

$$\lim_{a\to\infty} \int_4^a 1/x^{1/2}\, dx = \lim_{a\to\infty} \int_4^a x^{-1/2}\, dx$$

$$= \lim_{a\to\infty} 2x^{1/2} \Big[_4^{a^4} = \lim_{a\to\infty} 2a^{1/2} - 2(4)^{1/2}$$

But $a^{1/2}$ goes to infinity as $a\to\infty$. Therefore this integral DIVERGES.

Note: In the kind of integral of Example 2, if the exponent in the denominator is less than or equal to 1, the integral diverges. If the exponent is greater than 1, the integral would converge.

EXAMPLE 3 $\int_0^{\infty} \cos x\, dx = \lim_{a\to\infty} \int_0^a \cos x\, dx = \lim_{a\to\infty} \sin x \Big[_0^a$

$$= \lim_{a\to\infty} \sin a - \sin 0$$

This integral does not go to infinity. Yet it still diverges since, as a goes to infinity, sin (a) takes on every value between -1 and 1. The integral diverges because it doesn't go to 1 finite value.

If we have $\int_{-\infty}^{\infty} f(x)\, dx$, we break it up into 2 pieces, $\int_{-\infty}^{c} f(x)\, dx + \int_c^{\infty} f(x)\, dx$, where for convenience, c is often 0 but certainly does not have to be.

EXAMPLE 4 $\int_{-\infty}^{\infty} e^{-x}\, dx$

Whenever you have infinity at both ends, you should try to do the piece that diverges first. If you choose the piece that diverges, you do not have to do the other piece since the integral diverges (the whole integral diverges!!!). If the first piece converges, then you still must do the other piece. You sharp-eyed readers have probably spotted the fact that the negative infinity piece diverges since, roughly speaking, $e^{-(-\infty)}$ goes to infinity. Note $e^{-(\infty)}$ is 0.

Let us take a look at the other infinity kind of improper integral.

EXAMPLE 5 $\int_1^5 \frac{1\, dx}{(x-1)^{1/2}}$ **At $x = 1$, $f(x) = (x-1)^{-1/2}$ is infinite, the improper part**

$$= \lim_{a\to1^+} \int_a^5 (x-1)^{-1/2}\, dx = \lim_{a\to1^+} 2(x-1)^{1/2} \Big[_a^5 = \lim_{a\to1^+} 2(5-1)^{1/2} - 2(a-1)^{1/2}$$

$$= 4 - 0 = 4$$

This integral CONVERGES to 4.

Note: For this kind of improper integral, if the exponent in the denominator is less than 1, the integral converges; if the exponent is greater than or equal to 1, it diverges.

EXAMPLE 6

$$\int_2^5 \frac{1\,dx}{(x-3)^3}$$

First, note that this *is* an improper integral since f(3) is undefined. Second, most of the time, if 1 piece diverges, both diverge, so that it is not important which piece to choose first.

$$= \lim_{c\to 3^-} \int_2^c (x-3)^{-3}\,dx + \lim_{d\to 3^+} + \int_d^5 (x-3)^{-3}\,dx$$

$$= \lim_{c\to 3^-} -\frac{1}{2(x-3)^2}\Big[_2^c - \lim_{d\to 3^+} \frac{1}{2(x-3)^2}\Big[_d^5$$

$$= \lim_{c\to 3^-}\left[-\frac{1}{2(c-3)^2} + \frac{1}{2(2-3)^2}\right] + \lim_{d\to 3^+}\left[-\frac{1}{2(5-3)^2} + \frac{1}{2(d-3)^2}\right]$$

Each bracketed piece goes to infinity. The integral DIVERGES. However if you did the problem, you should calculate only *1* piece. Once it diverges, you need not do the other piece.

The next 2 are a couple of my most favorite examples. It makes you believe that mathematicians can do ANYTHING. It is not true, but the examples are extraordinary.

EXAMPLE 7 We will take an infinite area, rotate the region, and get a finite volume!!!!!!!!!

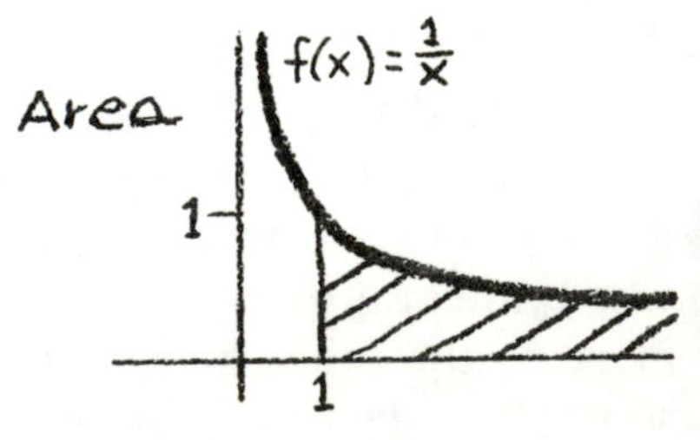

$$\int_1^\infty \frac{1}{x}\,dx = \lim_{a\to\infty} \int_1^a \frac{1}{x}\,dx$$

$$= \lim_{a\to\infty} \ln x \Big[_1^a = \lim_{a\to\infty} \ln a - \ln a \to \infty$$

Infinite area.

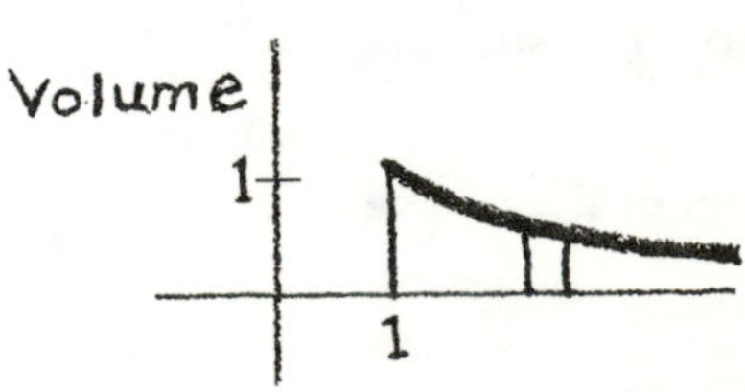

Volumes by sections. Much more on this is in Chapter 11 in this book.

$$\pi \int_1^\infty \left(\frac{1}{x}\right)^2 dx = \lim_{a\to\infty} \pi \int_1^a x^{-2}\,dx = \lim_{a\to\infty} \pi\left[-\frac{1}{x}\right]_1^a$$

$$= \lim_{a\to\infty} \pi\left[-\frac{1}{a} + \frac{1}{1}\right] = \pi$$

Amazing!!!!!!! Infinite area rotated gives finite volume.

This one will totally blow your mind. We will now take a finite region, rotate it, and get an infinite volume, which would seem impossible after the last example. It is NOT!!!!

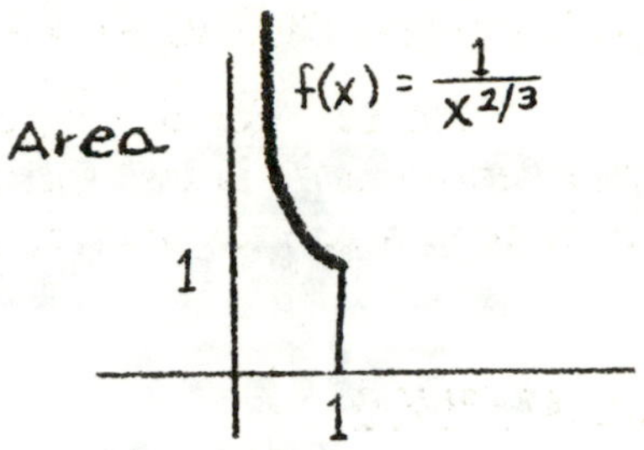

EXAMPLE 8 $\int_0^1 \frac{1}{x^{2/3}}\,dx = \lim_{a\to 0^+} \int_a^1 x^{-2/3}\,dx = \lim_{a\to 0^+} 3x^{1/3} \Big[_a^1$

$$= \lim_{a\to 0^+} (3 - 3a^{1/3}) = 3$$

Integral converges. Area is 3.

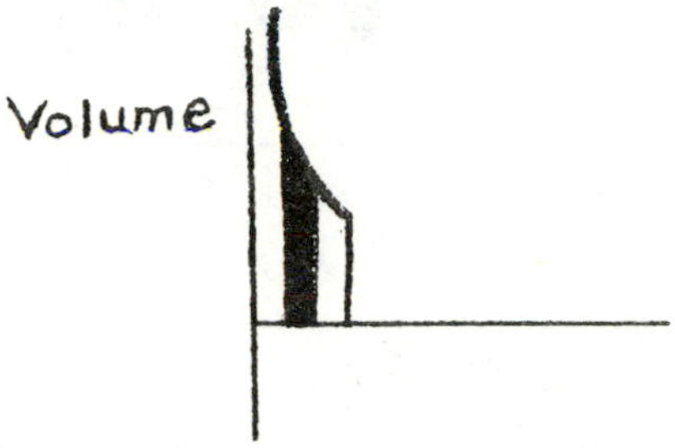

$$\pi \int_0^1 \left(\frac{1}{x^{2/3}}\right)^2 dx = \lim_{b\to 0^+} \pi \int_b^1 x^{-4/3}\,dx = \lim_{b\to 0^+} \pi\left(-\frac{3}{x^{1/3}}\right)\Big[_b^1$$

$$= \lim_{b\to 0^+} \pi\left(-3 + \frac{3}{b^{1/3}}\right) = \infty$$

Area

$f(x) = \frac{1}{x^{2/3}}$

1

1

Volume

Diverges!!!! Amazing!!!!!

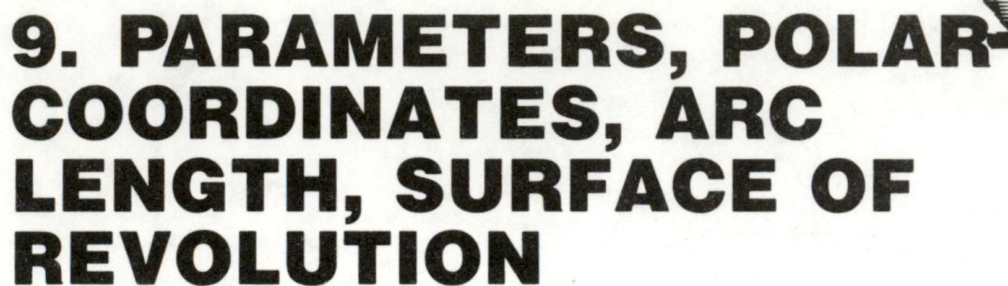

9. PARAMETERS, POLAR COORDINATES, ARC LENGTH, SURFACE OF REVOLUTION

If the x-y coordinate system were superior to all others, this chapter would not be necessary. However it is not. Parameters are variables which are introduced to make life easier.

EXAMPLE 1 $x = 2t$. $y = t^2/4 - 1$. $-2 \le t \le 4$.

We need to make a table with 3 columns: t,x,y. We will use only the x and y on the graph.

t	x	y
−2	−4	0
−1	−2	−3/4
0	0	−1
1	2	−3/4
2	4	0
3	6	5/4
4	8	3

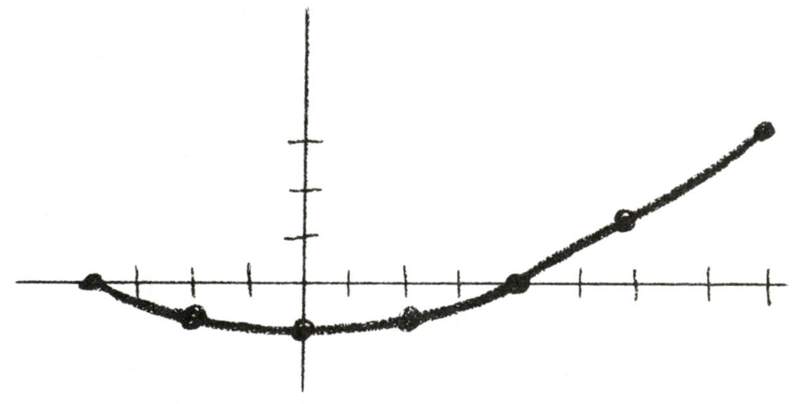

We can eliminate t. $x = 2t$. $t = x/2$. $y = (x/2)^2/4 - 1 = x^2/16 - 1$. It is clear we did not really need the parameter t in this case. However we did not want to take a too difficult example to show the use of a parameter.

EXAMPLE 2 Another common example is $x = \cos t$, $y = \sin t$. Again we don't need a parameter since $x^2 + y^2 = 1$.

EXAMPLE 3 Take a circle tangent to the x-axis at the origin. Put a dot on that point. Roll the circle on the x-axis. If we trace out the path of that dot we will get a curve called a CYCLOID. Since this curve is derived in most books, we will put the equations down and then draw the picture.

$$x = a(t - \sin t)$$

$$y = a(1 - \cos t)$$

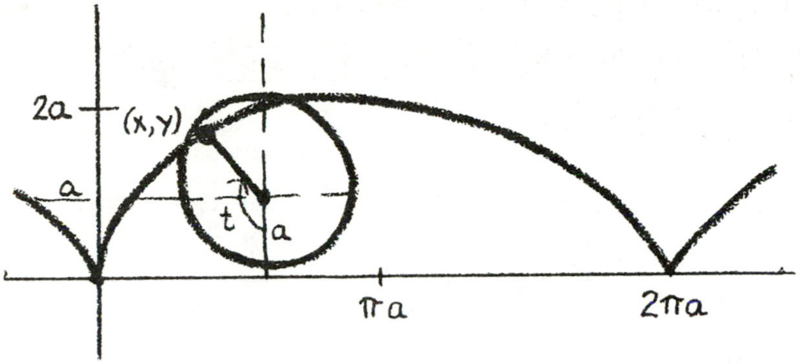

In order to eliminate t, we solve for t in the equation for y. $t = \cos^{-1}(1 - y/a)$. Therefore $x = a\{\cos^{-1}(1 - y/a) - \sin[\cos^{-1}(1 - y/a)]\}$. Pretty awful, isn't it?! For all practical purposes, this form is impossible to use. We really need parameters here!!!!

We wish to take derivatives using parameters. The first derivative will give us the slope and the second will tell its upness and downness.

EXAMPLE 4 $x = t^4 + 1$, $y = t^8 - 5$.

$$dy/dx = dy/xt \big/ dx/dt$$

$$dy/dt = 8t^7$$

$$dx/dt = 4t^3 \qquad \text{so} \qquad dy/dx = 8t^7/4t^3 = 2t^4$$

You might think d^2y/dx^2 is gotten by taking dy/dx and taking the derivative of top over the derivative of the bottom. You'd be wrong. You say, "Of course. You must use the quotient rule!" Again you'd be wrong. The second derivative is the derivative of the first derivative. So . . .

$$\frac{d^2y}{dx^2} = \frac{d(y')}{dx} = \frac{d(y')/dt}{dx/dt}$$

$$= \frac{d(2t^4)/dt}{d(t^4 + 1)/dt} = \frac{8t^3}{4t^3} = 2$$

Again in this particular example you could eliminate the t, but in the cycloid you really could not.

EXAMPLE 5 Cycloid $x = a(t - \sin t)$, $y = a(1 - \cos t)$.

$$dx/dt = a(1 - \cos t) \qquad dy/dt = a \sin t$$

$$dy/dx = (dy/dt)/(dx/dt) = (\sin t)/(1 - \cos t)$$

The a's cancel

$$dy'/dt = \frac{(1 - \cos t)(\cos t) - (\sin t)(\sin t)}{(1 - \cos t)^2} = \frac{\cos t - \cos^2 t - \sin^2 t}{(1 - \cos t)^2}$$

$$= \frac{\cos t - 1}{(1 - \cos t)^2} = \frac{-1}{1 - \cos t}$$

$$\frac{dy'}{dx} = \frac{dy'}{dt} \Big/ \frac{dx}{dt} = \frac{-1}{1 - \cos t} \Big/ a(1 - \cos t) = \frac{-1}{a(1 - \cos t)^2}$$

If you look at the picture of the cycloid, the second derivative shows the curve is always down since the second derivative is always negative ("a" pos.), except at multiples of 2π, where the curve comes to a point. The parameter is extremely useful here, as it always is when it is used.

Polar coordinates

In the past you should have had a teeny tiny bit of experience with polar coordinates, namely how to graph a point, say $(4, \pi/6)$. The following is a summary of all the things you should know on this topic.

Rule(s) 1 Relationship between (x, y) and (r, θ)

A. $r = (x^2 + y^2)^{1/2}$

B. θ = angle measured from positive x-axis, positive in a counterclockwise direction

C. $x = r \cos \theta$

D. $y = r \sin \theta$

Rule 2 Every point has an infinite number of representations.

$$(3, \pi/6) = (3, \pi/6 \pm 2n\pi) = (-3, 7\pi/6 \pm 2n\pi) \qquad n \text{ integer}$$

Rule(s) 3 A. $r = a$ is a circle center at the origin, radius a.

B. $\theta = c$ is a line through the origin.

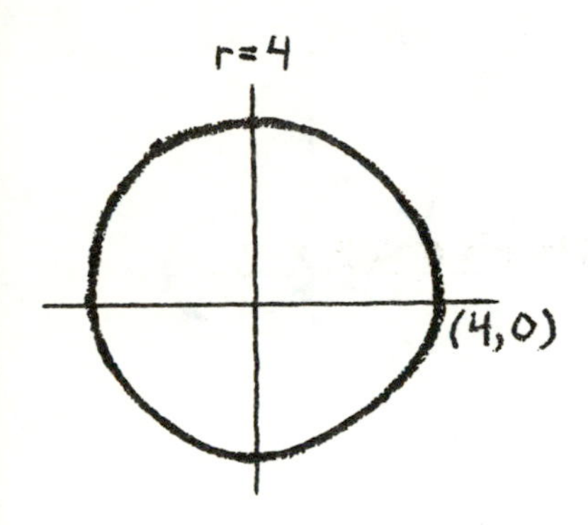

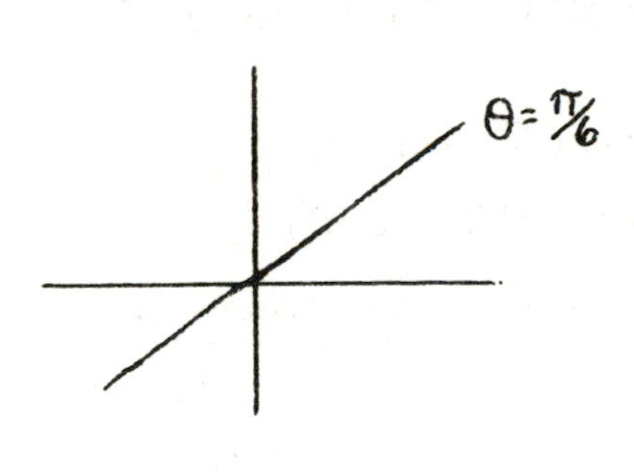

Rule(s) 4 Symmetries to aid polar curve sketching*

A. x-axis sym. If replacing (r,θ) by $(r,-\theta)$ or $(-r,\pi-\theta)$ gives the same or equivalent equation.

B. y-axis sym. If replacing (r,θ) by $(-r,-\theta)$ or $(r,\pi-\theta)$ gives the same or equivalent equation.

C. Origin sym. If replacing (r,θ) by $(-r,\theta)$ or $(r,\pi+\theta)$ gives the same or equivalent equation.

There are several curves that you should know by looking at the equation. It is always easier to sketch a curve if you know what it looks like before you start.

Rule 5 $r = a \sin \theta$

$r = a \cos \theta$

Circles tangent to origin, radius $|a|/2$, x- or y-axis is the axis of symmetry.

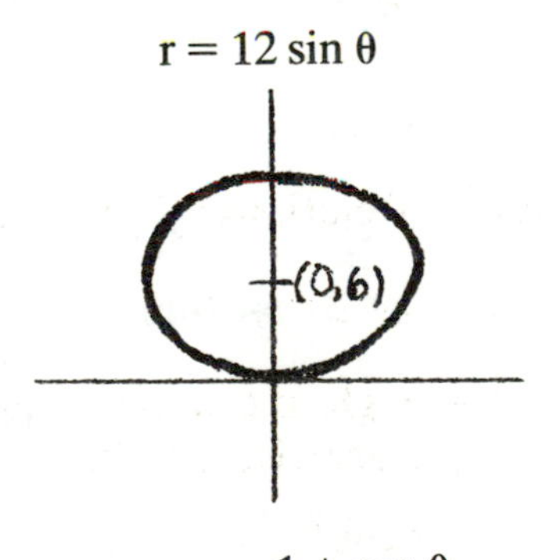

$r = 1 + \cos \theta$

Rule 6 Cardioid

$r = a(1 \pm \cos \theta)$

$r = a(1 \pm \sin \theta)$

*Two of the symmetries mean all 3 hold.

Rule 7 Rose–n positive odd

$r = a \sin n\theta$ or $a \cos n\theta$

n petals

$r = \cos 3\theta$, 3 petals

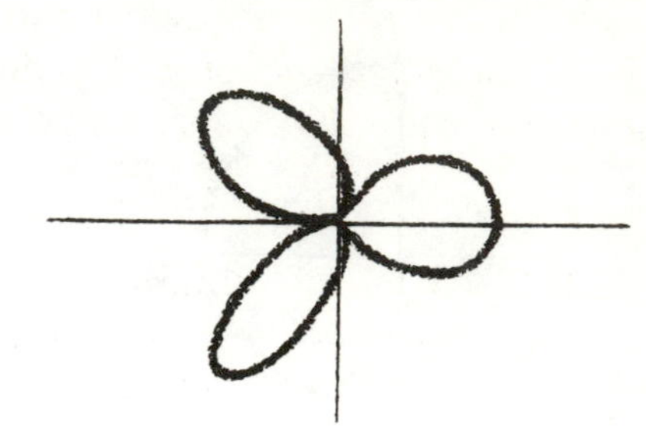

Rule 8 Rose–n positive even

$r = a \sin n\theta$ or $a \cos n\theta$

2n petals

$r = \sin 2\theta$, 4 petals

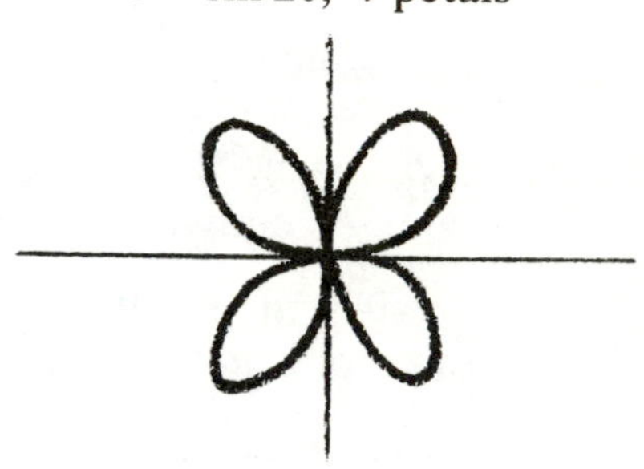

Rule 9 Lemniscate

$r^2 = \pm a \cos 2\theta$

$r^2 = \pm a \sin 2\theta$

$r^2 = \sin 2\theta$

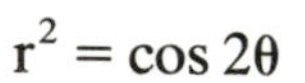

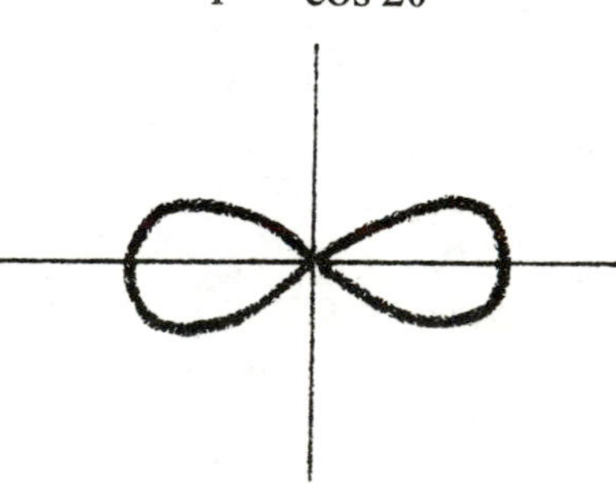

Rule 10 Spirals

$r = a\theta$ spiral of Archimedes

$r = ae^{b\theta}$ logarithmic spiral

$r = \theta$

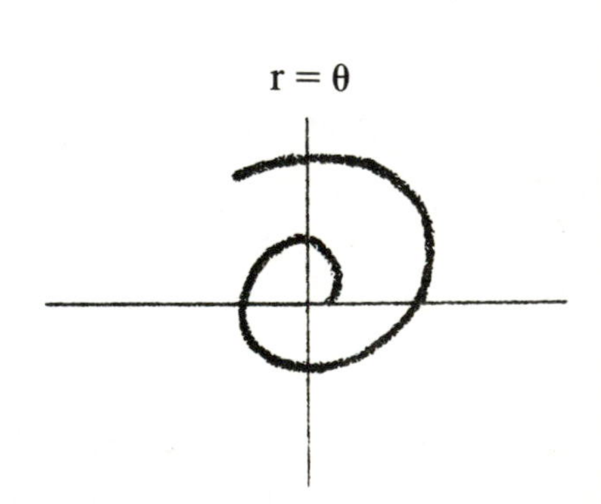

There are a few others, but these are the main ones. Let us graph a couple of these to show the technique.

EXAMPLE 6 Graph $r = 4(1 - \cos\theta)$.

θ	r
0	$4 - 4 = 0$
$\pi/6$	$4 - 2(3)^{1/2} = .6$
$\pi/4$	$4 - 2(2)^{1/2} = 1.2$
$\pi/3$	$4 - 2 = 2$
$\pi/2$	$4 - 0 = 4$
$2\pi/3$	$4 + 2 = 6$
$3\pi/4$	$4 + 2(2)^{1/2} = 6.8$
$5\pi/6$	$4 + 2(3)^{1/2} = 7.4$
π	$4 + 4 = 8$

Note: We have symmetry with respect to the x-axis. So we need only values of θ between 0 and π and reflect image in the x-axis. Note the chart. If you study the patterns, you should be able to do much of the table by sight. Also you know approx. value of $2^{1/2} = 1.4$ and $3^{1/2} = 1.7$.

Note: There is a short way to draw the previous graph (and other graphs) if you know what the graph is. Find the intercepts, 0, $\pi/2$, π, $3\pi/2$, and sketch the graph around them (which is exactly how I did the graph).

EXAMPLE 7 $r^2 = 9\cos 2\theta$

Symmetry is about x-axis, y-axis, and the origin. So we only need to graph the first quadrant, and then repeated reflectly the curve. However since we have 2θ, we need 3 columns—2θ, θ, and r—with only the last 2 on the graph.

2θ	θ	r
0	0	3
$\pi/6$	$\pi/12$	2.8
$\pi/4$	$\pi/8$	2.5
$\pi/3$	$\pi/6$	2.1
$\pi/2$	$\pi/4$	0

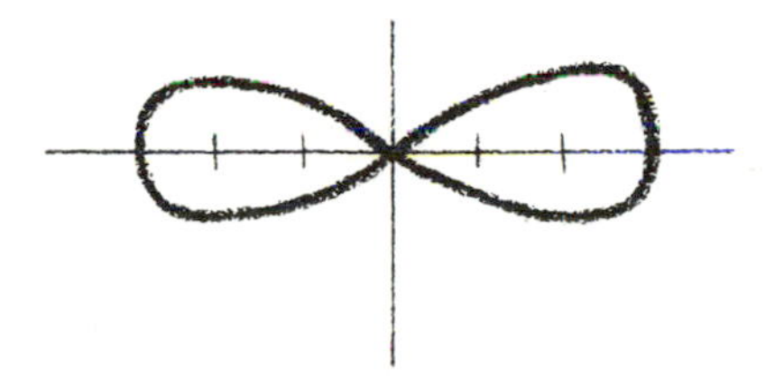

Although this problem can be done without a calculator if you can approximate well, it is convenient to use a calculator for this one.

We would now like to find the area in polar co-ordinates. In almost any calc book, you will find the area is $\frac{1}{2}\int_{\theta=\theta_1}^{\theta_2} r^2\, d\theta$.

EXAMPLE 8 Find the area of the cardioid $r = 2 + 2\cos\theta$.

As we have seen before, this curve is symmetric with respect to the x-axis. We can integrate from 0 to 2π or double the integral from 0 to π.

$$2\left(\tfrac{1}{2}\right)\int_0^{\pi} (2 + 2\cos\theta)^2\, d\theta = \int_0^{\pi} (4 + 8\cos\theta + 4\cos^2\theta)\, d\theta$$

$$= \int_0^{\pi} [4 + 8\cos\theta + 4(1 + \cos 2\theta)/2]\, d\theta$$

$$= \int_0^{\pi} (6 + 8\cos\theta + 2\cos 2\theta)\, d\theta$$

$$= 6\theta + 8\sin\theta + \sin 2\theta \Big[_0^{\pi} = 6\pi$$

EXAMPLE 9 Find area of lemniscate $r^2 = 9 \cos 2\theta$.

From Example 7 we have 4-quadrant symmetry. So we will find area in first quad. and multiply by 4.

$$\tfrac{1}{2} \int r^2 \, d\theta = 4(\tfrac{1}{2}) \int_0^{\pi/4} 9 \cos 2\theta \, d\theta = (2)(9) \; \frac{\sin 2\theta}{2} \Big[_0^{\pi/4} = 9$$

Here's a little trick about finding the area of a curve with petals.

EXAMPLE 10 Find the area of $r = 4 \sin 6\theta$.

This has 12 petals. We will find the area of 1 and multiply by 12. To find the area of 1 petal, we find the first 2 values $\sin 6\theta = 0$. $6\theta = 0$ or $\theta = 0$. $6\theta = \pi$. $\theta = \pi/6$. So the integral is $(12)\frac{1}{2} \int_0^{\pi/6} (4 \sin 6\theta)^2 \, d\theta$, etc.

Here's another trick.

EXAMPLE 11 Find the area of $r = 4 \cos 6\theta$.

We know we can slide the curve $y = f(x)$ +a units to the right by replacing x by $x - a$. In the same way we can rotate $r = f(\theta)$ through a counterclockwise angle $+\alpha$ by replacing θ by $\theta - \alpha$. Thus by rotating our curve by $15° = \pi/12$ radians, $r = 4 \cos 6(\theta - \pi/12) = 4 \cos (6\theta - \pi/2) = 4 \sin 6\theta$, which is exactly the curve in example 10!!!!! Now rotating a curve doesn't affect its area. So . . . the answer must be the same as in Example 10!!!!

Just like with x-y coordinates, we would like to find the area between 2 curves. In the case of x-y, it was top curve minus the bottom curve or right curve minus left curve. In polar coordinates, it is outside curve minus inside. The formula is $\frac{1}{2} \int_{\theta_1}^{\theta_2} (r^2_{outside} - r^2_{inside}) \, d\theta$ where θ_1 and θ_2 are the intersection of the 2 curves.

EXAMPLE 12 Find the area of the region bounded by the cardioid $r = 2 + 2 \cos \theta$ (inside this curve) and the circle $r = 3$ (outside this curve).

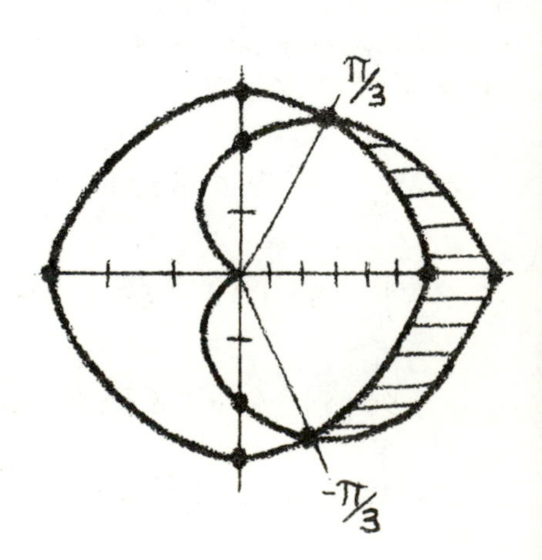

This figure is drawn to the right. Since the area is inside the cardioid, the cardioid is the OUTSIDE curve. Outside $r = 3$ makes the circle the INSIDE curve. In order to find the limits of the integral, we set the r's equal to each other. $3 = 2 + 2 \cos \theta$. $\cos \theta = \frac{1}{2}$. $\theta = \pm\pi/3$. Also there is symmetry about the x-axis. So we can double the integral from 0 to $\pi/3$.

$$2\left(\tfrac{1}{2}\right)\int_{\theta=0}^{\pi/3}(r_{\text{outside}}^2 - r_{\text{inside}}^2)\,d\theta = \int_0^{\pi/3}[(2+2\cos\theta)^2 - 3^2]\,d\theta$$

$$= \int_0^{\pi/3}[8\cos\theta + 4\cos^2\theta - 5]\,d\theta$$

$$= \int_0^{\pi/3}[8\cos\theta + (2+2\cos 2\theta) - 5]\,d\theta$$

$$= 8\sin\theta + \sin 2\theta - 3 = 8(3)^{1/2}/2 + 3^{1/2}/2 - 3(\pi/3) = \frac{9\sqrt{3}}{2} - \pi$$

There are 2 more problems that we can illustrate with this same diagram. Unfortunately this means I have to redraw this diagram twice more. Does anyone out there know a publisher?

EXAMPLE 13 Find the area outside the cardioid and inside the circle.

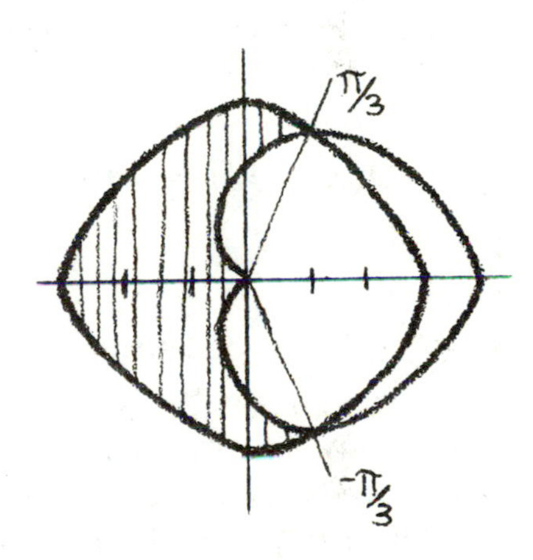

If it's inside the circle $r = 3$, $r = 3$ becomes the outside r. Outside the cardioid means $r = 2 + 2\cos\theta$ becomes the inside curve. Again we have x-axis symmetry. Again setting the curves equal, we get $\pm\pi/3$. The picture tells us twice the integral from $\pi/3$ to π. Since we did the last problem, we need not integrate twice since it is the same integral. We need only change all the signs.

$$2\left(\tfrac{1}{2}\right)\int_{\pi/3}^{\pi}[3^2 - (2+2\cos\theta)^2]\,d\theta = 3\theta - 8\sin\theta - \sin 2\theta\Big[_{\pi/3}^{\pi}$$

$$= [3\pi - 8\sin\pi - \sin 2\pi] - [\pi - 8\sin\pi/3 - \sin 2\pi/3]$$

$$= 2\pi + \frac{9\sqrt{3}}{2}$$

EXAMPLE 14 Find the area inside both the circle and the cardioid.

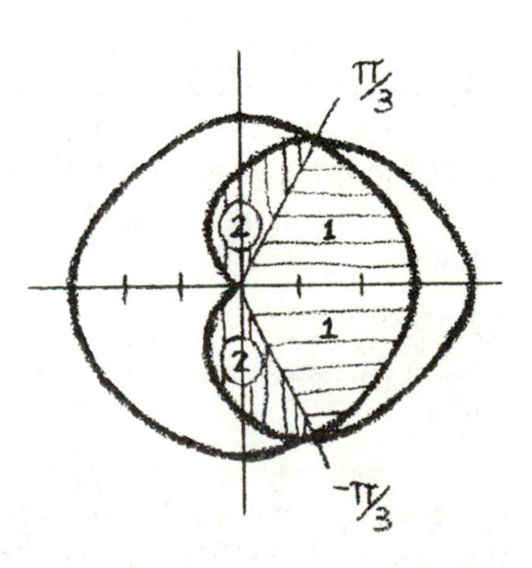

Now we know your instructor would never give such a problem, but. . . . This is a split region at the rays $\theta = \pm\pi/3$. Region 1 is a sector of the circle. Region 2 is the "top" of the heart, the cardioid. So we get

$$2\left(\tfrac{1}{2}\right)\int_0^{\pi/3}3^2\,d\theta + 2\left(\tfrac{1}{2}\right)\int_{\pi/3}^{\pi}(2+2\cos\theta)^2\,d\theta$$

$$= \int_0^{\pi/3}9\,d\theta + \int_{\pi/3}^{\pi}(6 + 8\cos\theta + 2\cos 2\theta)\,d\theta$$

$$= 9\theta\Big[_0^{\pi/3} + 6\theta + 8\sin\theta + \sin 2\theta\Big[_{\pi/3}^{\pi}$$

$$= 3\pi + 4\pi - \frac{9\sqrt{3}}{2} = 7\pi - \frac{9\sqrt{3}}{2}$$

Please note that polar coordinates are really not hard. It is just many instructors rush through this section. This makes this section appear difficult. It really is not.

ARC LENGTH AND SURFACE AREA OF REVOLUTION

The derivations are in most books. All that remains is memorizing the formulas and applying them.

Arc length:

$$\int_a^b [1 + f'(x)^2]^{1/2}\,dx \qquad \text{or} \qquad \int_c^d [1 + f'(y)^2]^{1/2}\,dy$$

or, in parameters,

$$\int_{t=t_1}^{t=t_2} [(dx/dt)^2 + (dy/dt)^2]^{1/2}\,dt \qquad \text{or} \qquad \int_{\theta=\theta_1}^{\theta=\theta_2} [f(\theta)^2 + f'(\theta)^2]^{1/2}\,d\theta$$

(curve is $r = f(\theta)$)

Surface of revolution about x-axis (distance is y)

$$\int_a^b 2\pi y[1 + (y')^2]^{1/2}\,dx \qquad \text{or} \qquad \int_c^d 2\pi y[1 + (x')^2]^{1/2}\,dy$$

Parametric:

$$\int_{t=t_1}^{t=t_2} 2\pi y(t)[(dx/dt)^2 + (dy/dt)^2]^{1/2}\,dt$$

Polar form:

$$\int_{\theta=\theta_1}^{\theta=\theta_2} 2\pi f(\theta)\sin\theta\,[f(\theta)^2 + f'(\theta)^2]^{1/2}\,d\theta$$

Note if about y-axis, x replaces y and $\theta \sin\theta$ is replaced by $\theta\cos\theta$.

Let us do 1 simple example, 1 clever example, and 1 known result.

EXAMPLE 15 $x = f(y) = (2/3)y^{3/2}$, arc length $0 \le y \le 1$.

$$f'(y) = y^{1/2} \qquad \text{so} \qquad [1 + f'(y)^2]^{1/2} = (1 + y)^{1/2}$$

$$\text{Arc length} = \int_0^1 (1 + y)^{1/2}\,dy = (2/3)(1 + y)^{3/2}\Big[_0^1 = (2/3)(2^{3/2} - 1)$$

EXAMPLE 16 (clever, very clever) $y = f(x) = (1/3)x^3 + 1/(4x)$, arc length $1 \le x \le 2$.

$$f'(x) = x^2 - (1/4x^2) \qquad f'(x)^2 = x^4 - \tfrac{1}{2} + 1/16x^4 = (x^2 - 1/4x^2)^2$$

$$+1 = +1$$

$$1 + f'(x)^2 = x^4 + \tfrac{1}{2} + 1/16x^4 = (x^2 + 1/4x^2)^2$$

$$[1 + f'(x)^2]^{1/2} = x^2 + 1/4x^2$$

Pretty clever???? Not when you have to see this for yourself it isn't.

$$\text{Arc length} = \int_1^2 (x^2 + 1/4x^2)\,dx = x^3/3 - 1/4x \Big[_1^2$$

$$= (8/3 - \tfrac{1}{2}) - (1/3 - \tfrac{1}{4}) = 25/12$$

EXAMPLE 17 Find the surface area of a sphere of radius a.

A sphere is a semicircle rotated about the x-axis (or y-axis). $x = a\cos t$. $dx/dt = -a \sin t$. $y = a \sin t$. $dy/dt = a \cos t$.

$$SA = 2\pi y[(dx/dt)^2 + (dy/dt)^2]^{1/2}\,dt$$

$$= 2\pi a \sin t[(-a \sin t)^2 + (a \cos t)^2]^{1/2}\,dt$$

Under the square root sign simplifies to a^2. So we get . . .

$$SA = 2\pi a^2 \int_0^{\pi} \sin t\,dt = 2\pi a^2(-\cos t) \Big[_0^{\pi} = 2\pi a^2[-(-1) + 1]$$

Again we haven't been lied to all these years. The surface area of a sphere is really $\boxed{4\pi a^2}$. . . .

10. WORK, WORK, WORK

This topic is usually presented in a physics book and scares everyone to death. If done my way, I'm pretty sure it won't bother you again. Work is defined as force times distance. However if the force is a function of distance, theory tells us that work is the integral of F(x) dx. We give the usual examples: springs and a couple on pumping water over the top.

EXAMPLE 1 It takes a force of 20 pounds to stretch a 7-foot spring to 11 feet. How much work to pull out the spring from 13 feet to 27 feet?

For a spring force $F = kx$. $F = 20$, $x = 11 - 7 = 4$, the distance the spring is stretched from its natural length. So $k = 5$ and $F = 5x$.

$$\text{Work} = \int kx\,dx \qquad \text{lower limit} = 13 - 7 = 6,\ \text{upper limit } 27 - 7 = 20$$

$$\text{Work} = \int_6^{20} 5x\,dx = (5x^2/2)\Big|_6^{20} = 5(20)^2/2 - 5(6^2)/2$$

$$= 1000 - 90 = 910 \text{ foot-pounds}$$

EXAMPLE 2 How much work is done to pump the water out of a full cylindrical can radius 10 feet, height 20 feet—if the water is to be pumped over the top?

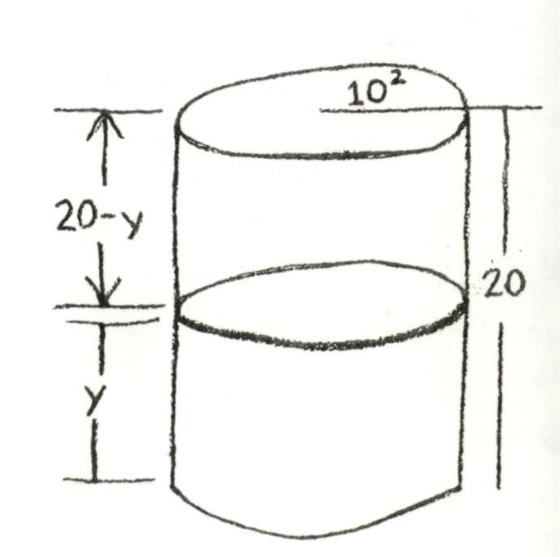

W = density (weight/volume) × volume × height over which the water is pumped. Each section of water pumped is a thin cylinder $V = \pi(10)^2\,dy$. Height pumped (look at picture) is $20 - y$. Density for water is approx. 62.5 pounds per cubic foot.

$$\int_0^{20} 62.5\pi(10)^2(20 - y)\,dy = 6250\pi(20y - y^2/2)\Big|_0^{20}$$

$$= 1{,}250{,}000 \text{ foot-pounds}$$

Note 1: If the outlet were 17 feet over the can, we would have $37 - y$.

Note 2: If the can was three-fourths full, integral limits would be 0 to 15.

Note 3: If we did the same problem with a box, cross section would be a thin sheet with length and width constant and height dy.

EXAMPLE 3 How much work is done to pump water out of a cone, diameter 22 feet, height 10 feet, if the outlet is 7 feet over the top of the cone and the cone is filled 2 feet deep?

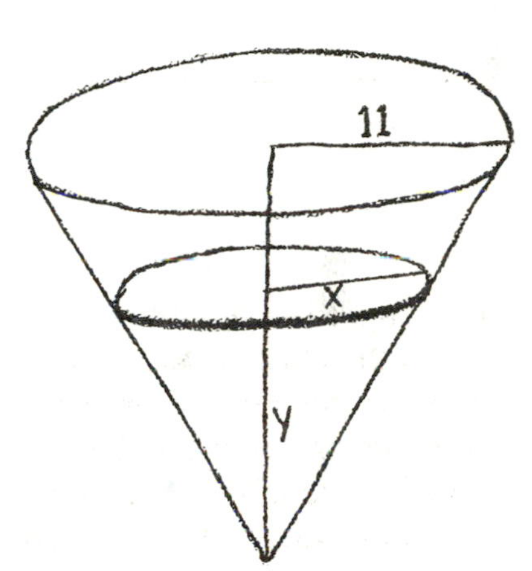

First a slight trick. Diameter is 22 so the radius is 11. Next note the cross section is again a thin cylinder, but this time the radius changes. $V = \pi r^2 h = \pi x^2\, dy$. We must see a similar triangle $x/y = 11/10$ or $x = 11y/10$. The height of the pipe makes the pumping distance $17 - y$.

We leave the integration to you. $\int_0^2 62.5\pi(\frac{11}{10}y)^2(17 - y)\, dy$.

Note: A trough is a similar problem except the horizontal cross section is a rectangular sheet whose width changes, but whose length (length of the trough) remains the same. . . .

Also note similar triangles, just like in the cone.

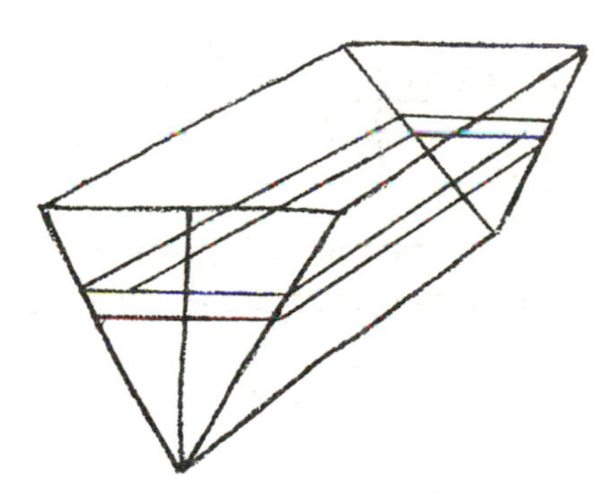

11. VOLUMES BY ROTATION AND SECTION

This chapter is the first of two that are reprinted from my first CALC HELPER, the beginnings of this course. The material in this chapter is in many Calc I courses. However, at my school it is in Calc II. The material in the next chapter, Conic Sections—Circle, Ellipse, Parabola, Hyperbola, appears in either Calc I or Calc II, depending on the book and the course. In my school it is in the non-math-major Calc I and math-major Calc II. Therefore, it rightly should be in both books.

The first topic is to find volumes of rotations. This is very visual. If you see the picture, the volume is easy. If not, this topic is very hard.

Imagine a perfectly formed apple with a line through the middle from top to bottom. We can find the volume 2 different ways. One way is by making slices perpendicular to the line (axis). (We will do the other way later with an onion.) Each slice is a disc, a thin cylinder. Its volume is $\pi r^2 h$, where h is very small. If we add up all the discs, taking the limits properly we get the volume.

We will take the same region in 6 different problems, rotating this region differently 6 times and getting 6 different volumes.

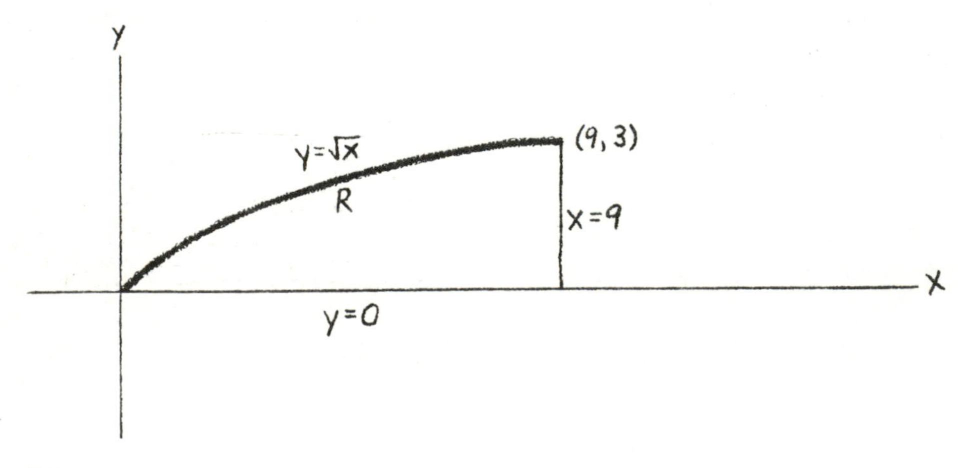

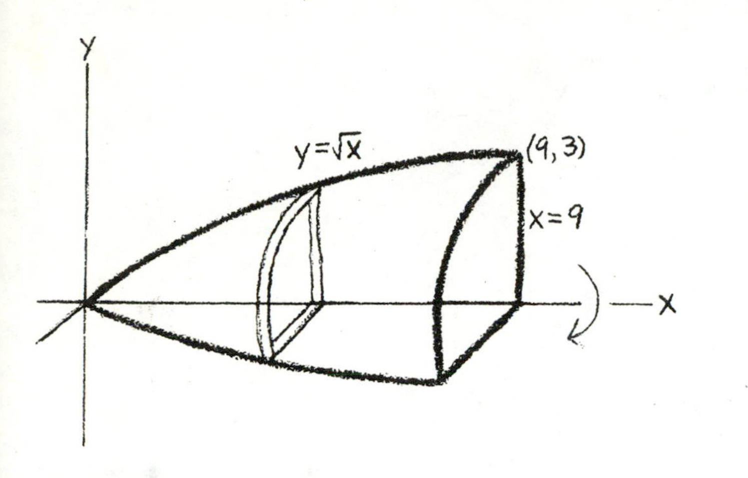

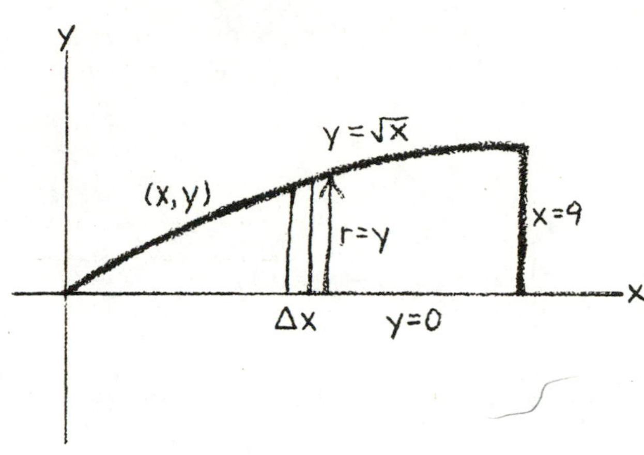

EXAMPLE 5 Find the volume if the region R is rotated about the x-axis.

The volume of each disc is $\pi r^2 h$. $h = \Delta x$. $r = y$. So $r^2 = y^2 = x$, x goes from 0 to 9.

$$V = \int_0^9 \pi r^2\, dh = \pi \int_{x=0}^9 x\, dx = \frac{\pi x^2}{2}\Big[_0^9 = \frac{81\pi}{2}$$

The integrals are almost always easy. Once you understand the picture, all will be easy. But it takes most people time, studying the pictures.

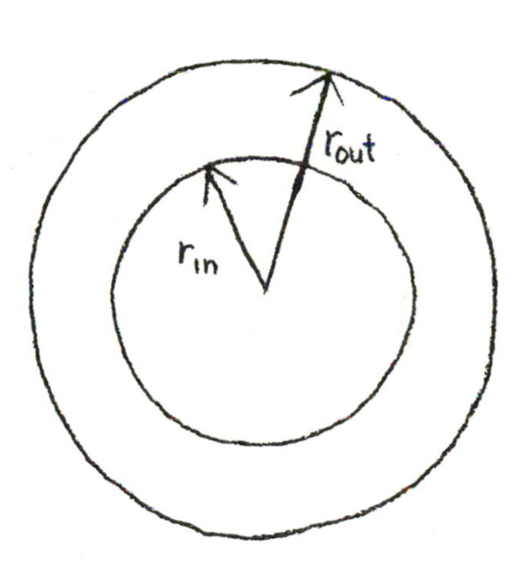

Let us get back to our apple. Suppose we core our apple. When we take slices perpendicular to the axis, we get a ring. The area of a ring is the area of the outside minus the area of the inside.

The volume of each disc is $(\pi r^2_{\text{outside}} - \pi r^2_{\text{inside}})h$. Again h is small.

EXAMPLE 6 Find the volume if our region is rotated about the y-axis.

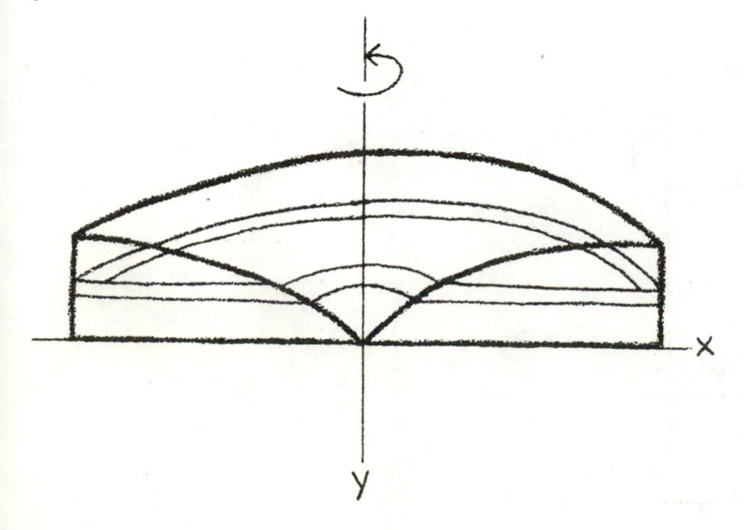

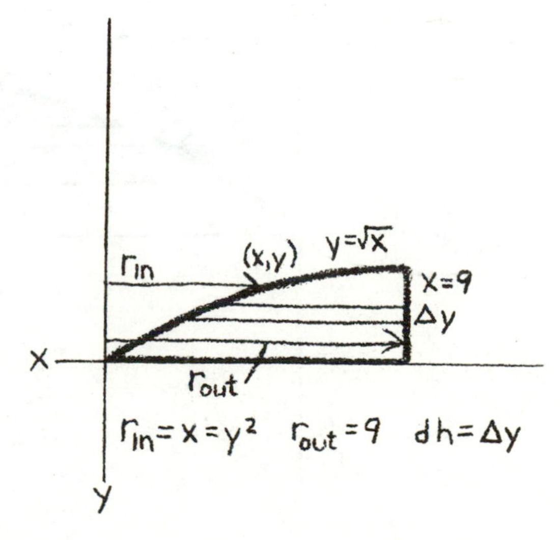

As you rotate this region there is a hole. Outside radius is always 9 and the inside radius is always the x value. But $x = y^2$. $r^2 = x^2 = y^4$.

$$V = \int_{y=0}^{3} (\pi r^2_{out} - \pi r^2_{in})\, dh = \pi \int_0^3 (9^2 - y^4)\, dy$$

$$= \pi\left[81y - \frac{y^5}{5}\right]_0^3 = \pi\left(81(3) - \frac{3^5}{5}\right) = \frac{972}{5}\pi$$

EXAMPLE 7 Find the volume of our glorious region if it is rotated about the line $x = 9$.

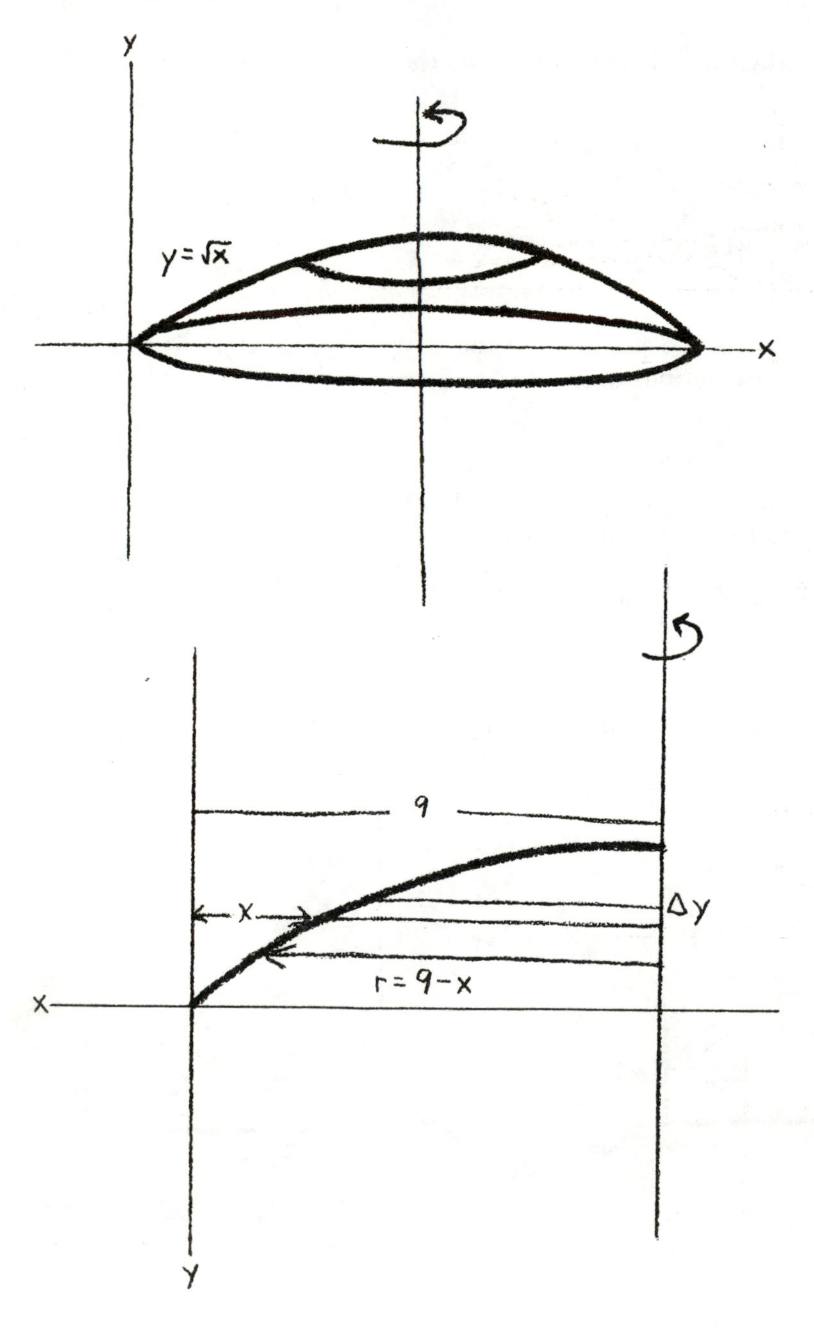

Notice that when we rotate the region about $x = 9$, there is no hole. $V_{sect} = \pi r^2 h$. $r = 9 - x = 9 - y^2$. $r^2 = 81 - 18y^2 + y^4$. $h = \Delta y$.

$$V = \pi \int_0^3 (81 - 18y^2 + y^4)\,dy = 81y - 6y^3 + \frac{y^5}{5}\Big[_0^3 \pi$$

$$= \pi\left[81(3) - 6(3)^3 + \frac{3^5}{5} - 0\right] = \frac{648\pi}{5}$$

EXAMPLE 8 Find the volume if the same region R is rotated about the line $x = -1$.

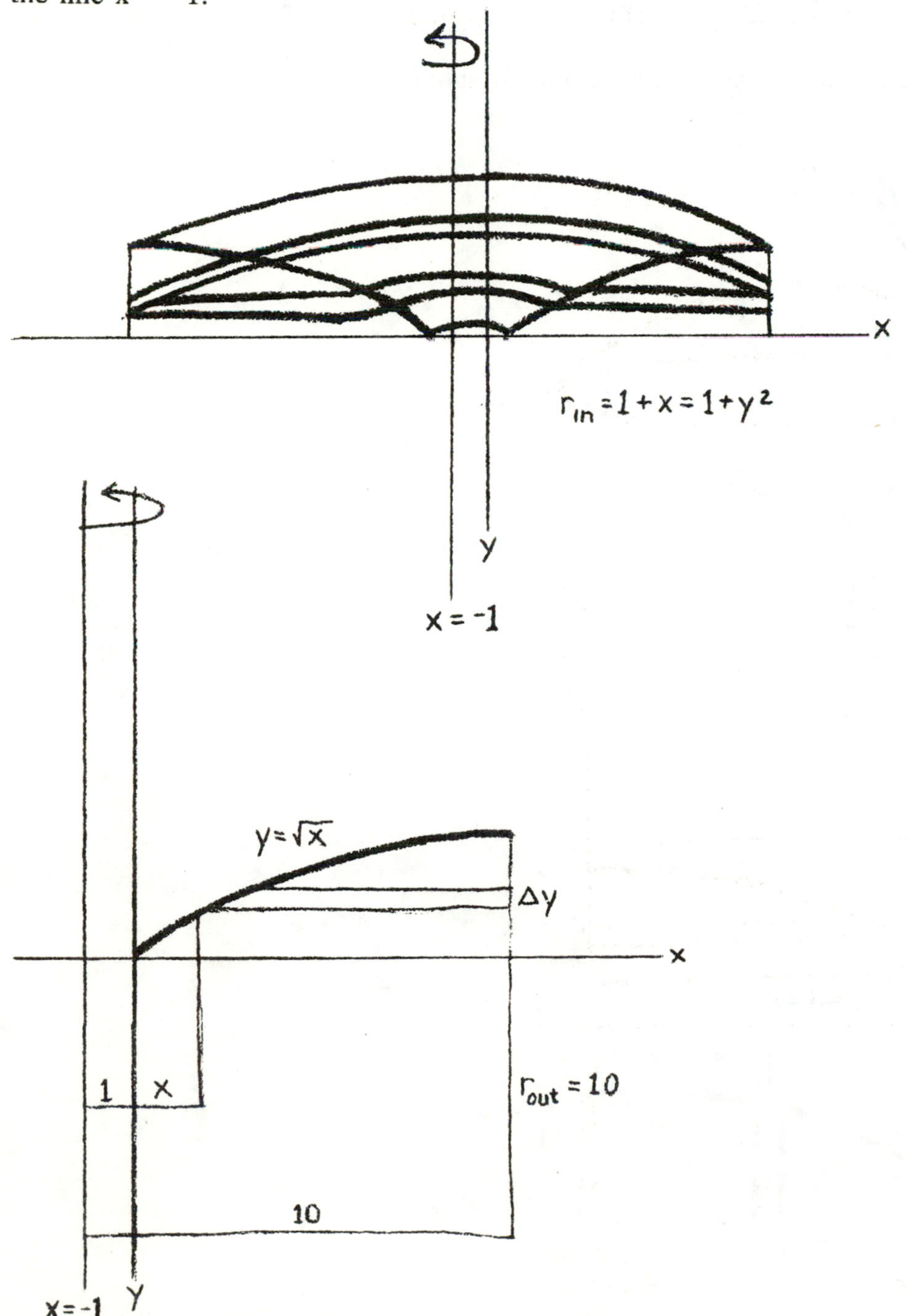

$V_{section} = \pi(r^2_{out} - r^2_{in})h$. $r_{out} = 9 + 1 = 10$. $r_{in} = 1 + x = 1 + y^2$. $r^2 = 1 + 2y^2 + y^4$. $h = \Delta y$.

$$V = \int_0^3 (\pi r^2_{out} - \pi r^2_{in})\,dy = \pi \int_0^3 [10^2 - (1 + 2y^2 + y^4)]\,dy$$

$$= \pi \int_0^3 (99 - 2y^2 - y^4)\,dy = 99y - \frac{2y^3}{3} - \frac{y^5}{5}\Big[_0^3 \pi$$

$$= \pi\left[99(3) - \frac{2(3)}{3} - \frac{3^5}{5} - 0\right] = \frac{1152\pi}{5}$$

EXAMPLE 9 Find the volume if our beloved region is rotated about $y = 5$.

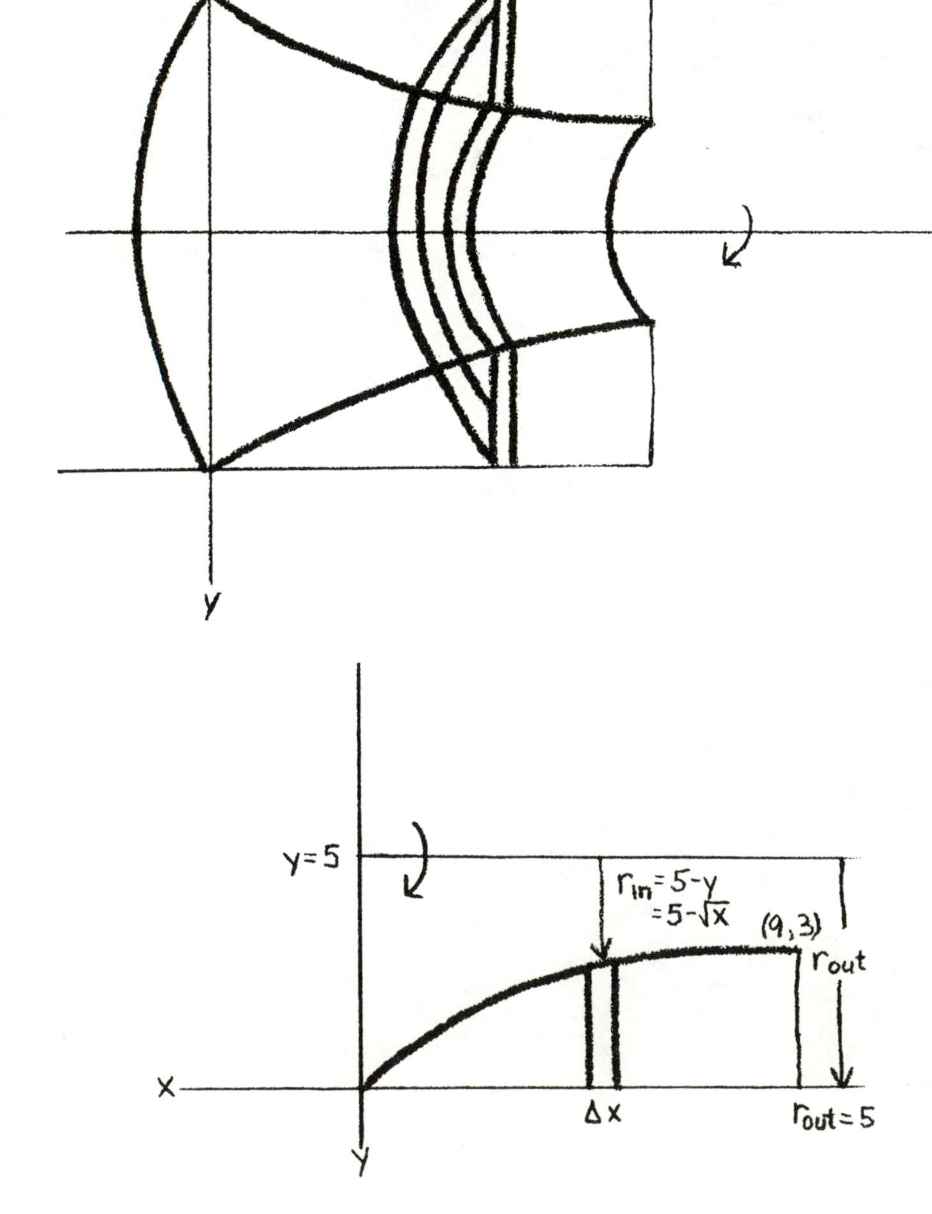

$V_{section} = \pi(r^2_{out} - r^2_{in})h$. $r_{out} = 5$. $r_{in} = 5 - y = 5 - x^{1/2}$. $r^2_{in} = 25 - 10x^{1/2} + x$. $h = \Delta x$.

$$V = \int_0^9 (\pi r^2_{out} - \pi r^2_{in})\,dx = \pi \int_0^9 [25 - (25 - 10x^{1/2} + x)]\,dx$$

$$= \pi \int_0^9 (10x^{1/2} - x)\,dx = \frac{20}{3}x^{3/2} - \frac{x^2}{2}\Big[_0^9 \pi$$

$$= \pi\left(\frac{20}{3}9^{3/2} - \frac{9^2}{2} - 0\right) = 139.5\pi$$

EXAMPLE 10 And for our final attraction we will take the same region and rotate it about the line $y = -2$.

$V_{sect} = (r^2_{out} - r^2_{in})h$. $r_{in} = 2$, $r_{out} = 2 + y = 2 + x^{1/2}$. $r^2_{out} = 4 + 4x^{1/2} + x$.

$$V = \int_0^9 (\pi r^2_{out} - \pi r^2_{in})\,dx = \pi \int_0^9 [(4 + 4x^{1/2} + x) - 4]\,dx$$

$$= \pi \int_0^9 (4x^{1/2} + x)\,dx = \frac{8}{3}x^{3/2} + \frac{x^2}{2}\Big[_0^9 \pi$$

$$= \pi\left(\frac{8}{3} \cdot 9^{3/2} + \frac{9^2}{2}\right) = \frac{225\pi}{2}$$

The next kind of volumes we will consider are rotations again, but we will do it a different way. Think about an onion with each layer a whole piece. We will add up layer by layer until we get a volume. We will add up cylindrical shells, tall cylindrical shells. We will see one of these shells, a general picture, and two examples.

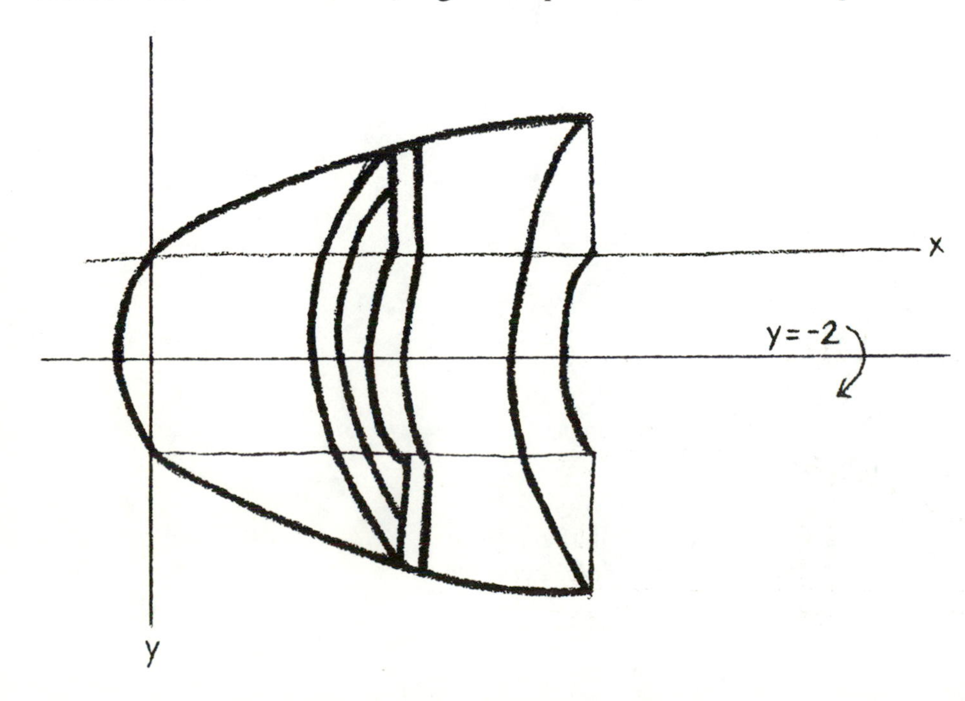

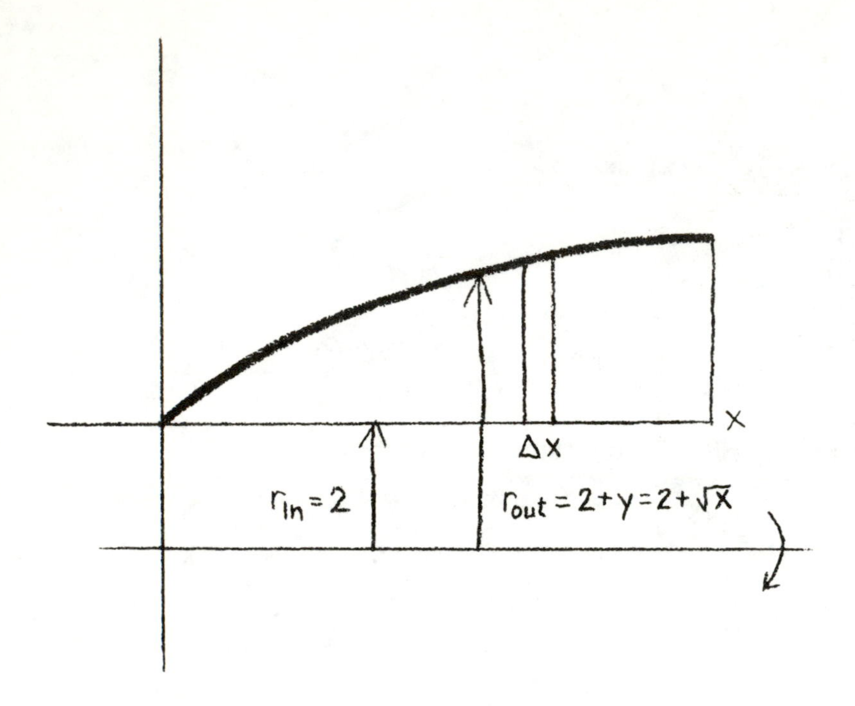

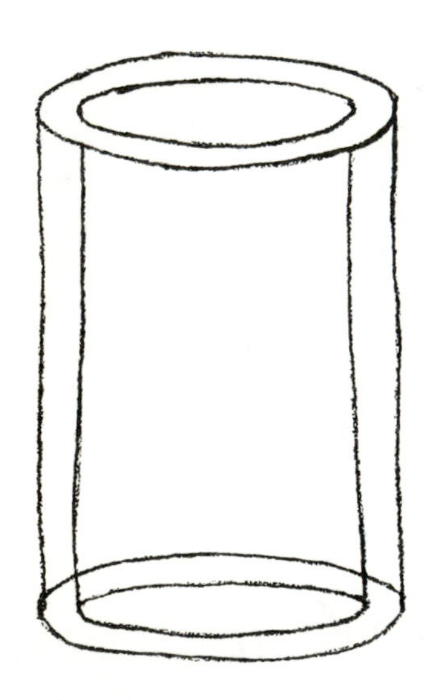

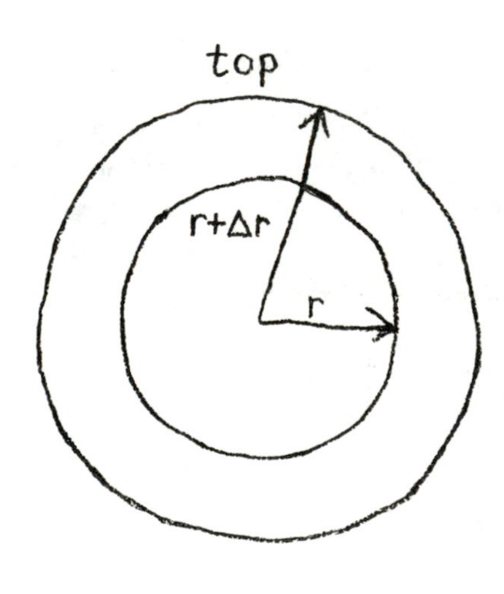

$$V_{shell} = \pi r^2_{out}h - \pi r^2_{in}h = \pi h(r^2_{out} - r^2_{in})$$

$$= \pi h(r_{out} + r_{in})(r_{out} - r_{in}) = 2\pi h\left(\frac{r_{out} + r_{in}}{2}\right)\Delta r$$

$$= 2\pi \cdot \text{height} \cdot \text{average radius} \cdot \text{thickness}$$

This is an example of a rotation about the y-axis of the region bounded by $y = f(x)$, $y = 0$, $x = a$, $x = b$. *Notice the axis of the cylinder is the axis which the curve is rotated around.*

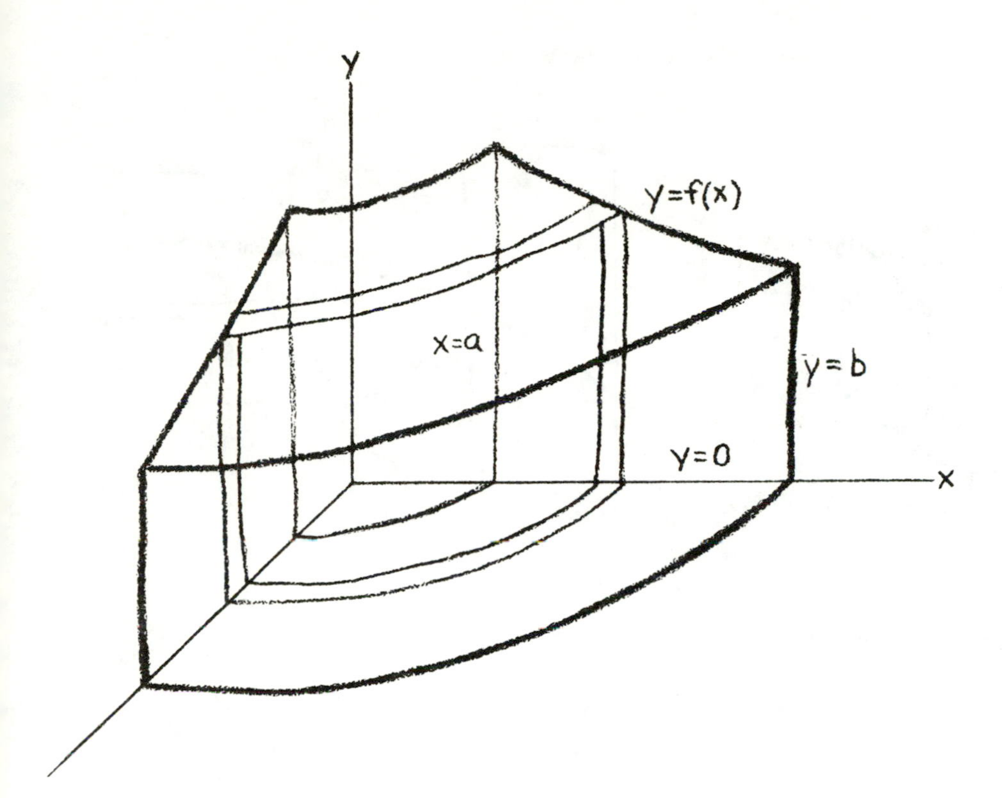

EXAMPLE 11 Let us try to do the same region R we had before and rotate it about the y-axis. We take the same region for 2 reasons: (1) you do not have to worry about different curves, and (2) we would like to show the answers are the same.

$$V = \int 2\pi \cdot \text{average radius} \cdot \text{average height} \cdot \text{thickness}$$

$$= 2\pi \int_0^9 x \cdot x^{1/2}\, dx = 2\pi \int_0^9 x^{3/2}\, dx = \frac{4}{5} x^{5/2} \Big[_0^9 \pi$$

$$= \frac{4\pi}{5}(9^{5/2} - 0) = \frac{972\pi}{5}$$

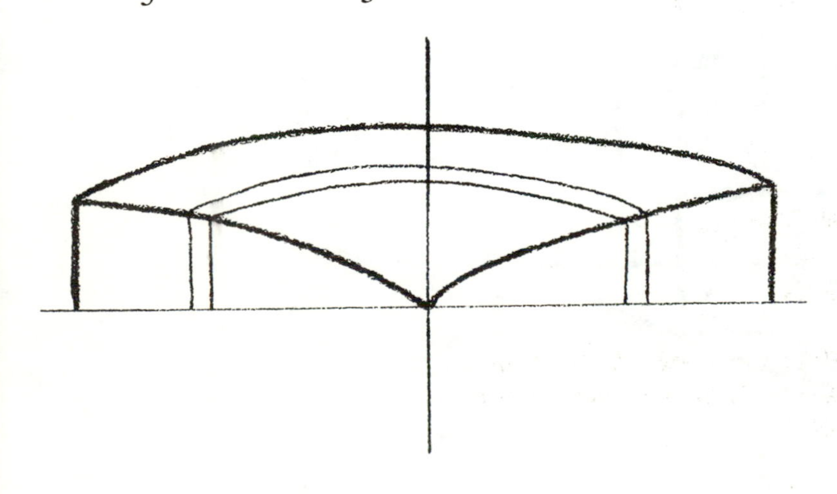

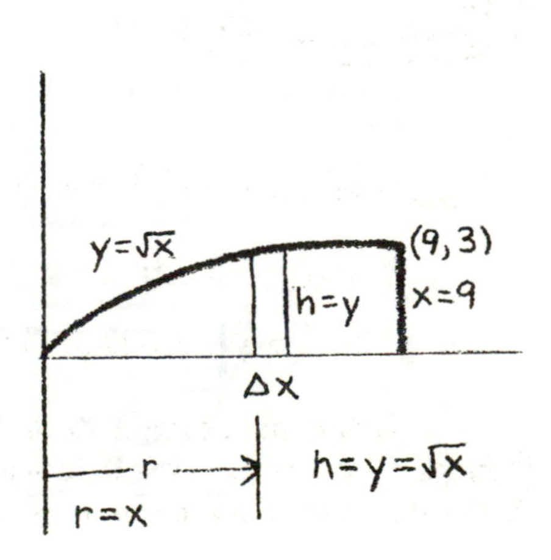

EXAMPLE 12 We will do our region one last time about the x-axis.

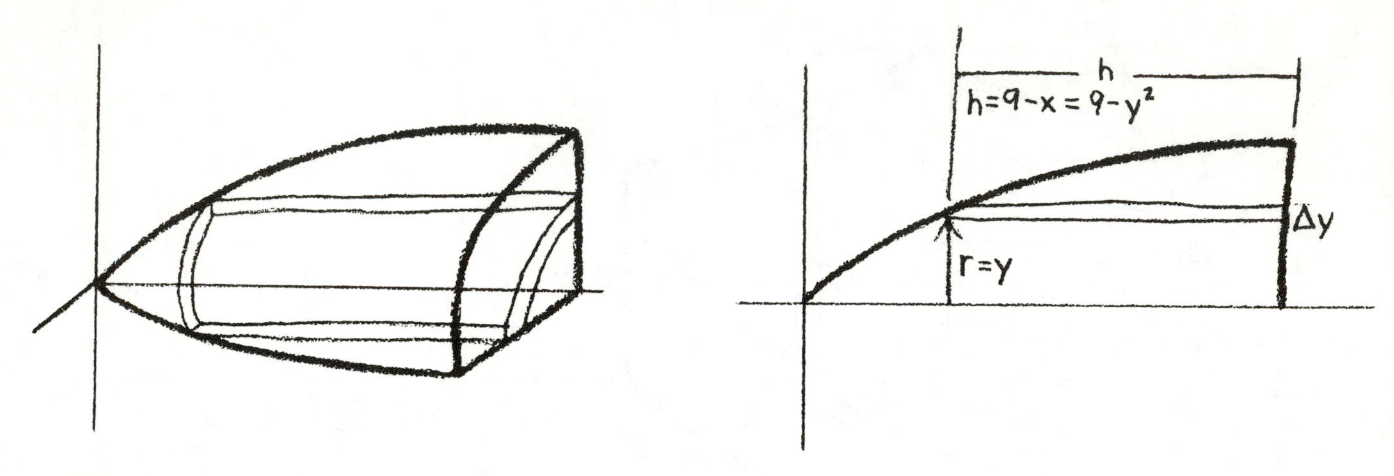

$$V = 2\pi \int_{y=0}^{3} \text{average radius} \cdot \text{average height} \cdot \text{thickness} = 2\pi \int_0^3 y(9 - y^2)\,dy$$

$$= 2\pi \int_0^3 (9y - y^3)\,dy = \frac{9y^2}{2} - \frac{y^4}{4} \Big[_0^3 2\pi = 2\pi\left[\frac{9(3)^2}{2} - \frac{3^4}{21}\right] = \frac{81\pi}{2}$$

The last part of this chapter is volume by sections. The sections are not circles but other shapes. Examples are given.

EXAMPLE 13 Find the volume of the following figure: Base $y = \frac{1}{4}x^2$ in the x-y plane. Sections perpendicular to the y-axis are rectangles with height 1/3 the base.

$A = bh = b(1/3b) = b^2/3 \qquad b = x - (-x) = 2x$

$y = \frac{1}{4}x^2 \qquad x^2 = 4y \qquad x = 2y^{1/2} \qquad 2x = 4y^{1/2} = b$

$A = (4y^{1/2})^2/3 = 16y/3$

Volume of each section $= 16y/3\ \Delta y$

$$V = \int_0^4 \frac{16y}{3}\,dy = \frac{8y^2}{3} \Big[_0^4 = \frac{8(4)^2}{3} - \frac{8(0)^2}{3} = \frac{128}{3}$$

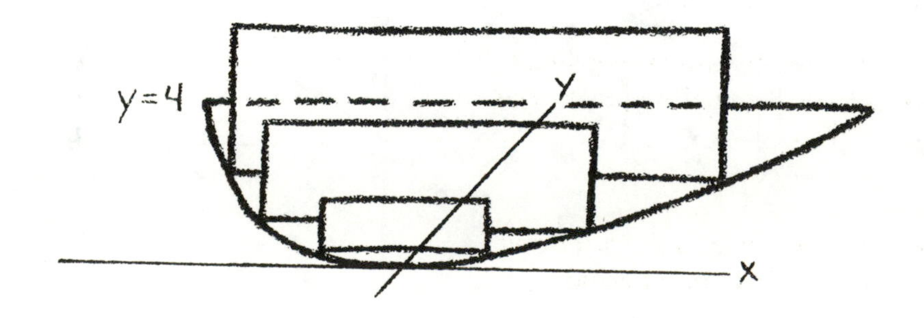

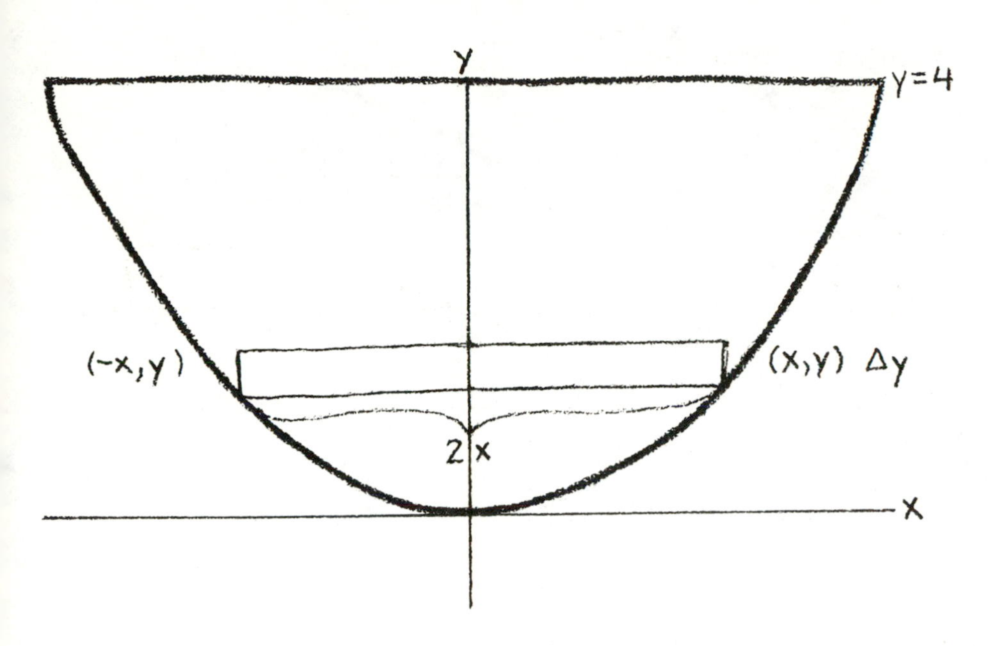

EXAMPLE 14 Base $x^2 + y^2 = 9$. Sections perpendicular to the x-axis are equilateral triangles. Find the volume.

$$\textbf{area} = \frac{s^2\sqrt{3}}{4}$$

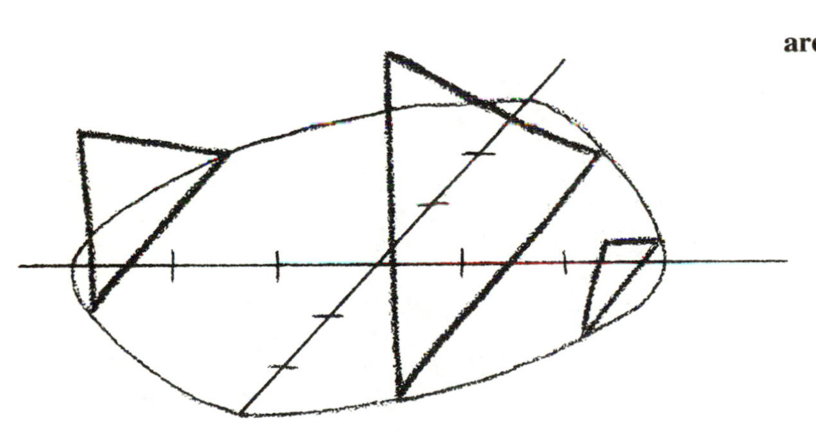

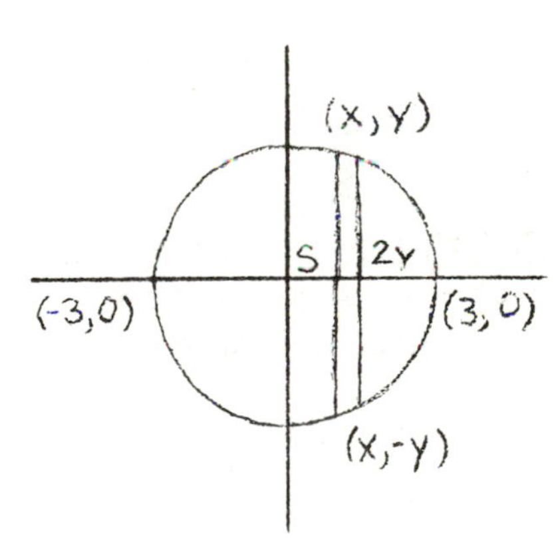

$$\int_{-3}^{3} A(x)\,dx \qquad A = \frac{s^2\sqrt{3}}{4} = \frac{(2y)^2\sqrt{3}}{4} = \sqrt{3}y^2 = \sqrt{3}[9 - x^2]$$

$$\sqrt{3}\int_{-3}^{3} (9 - x^2)\,dx = \sqrt{3}\left[9x - \frac{x^3}{3}\right]_{-3}^{3} = \sqrt{3}(18 - (-18)) = 36\sqrt{3}$$

Even though the base is a circle, we are adding triangular slices. Since the area of a triangle does not involve pi, neither does our volume.

EXAMPLE 15 Base bounded by $y = x^{1/2}$, $x = 9$, x axis. Sections perpendicular to the x-axis are semicircles. Find the volume.

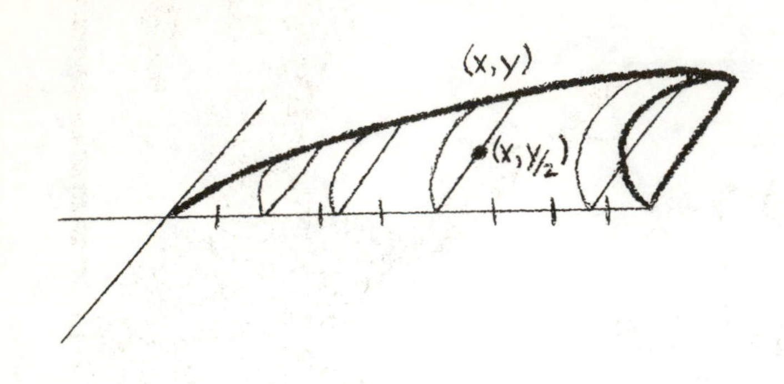

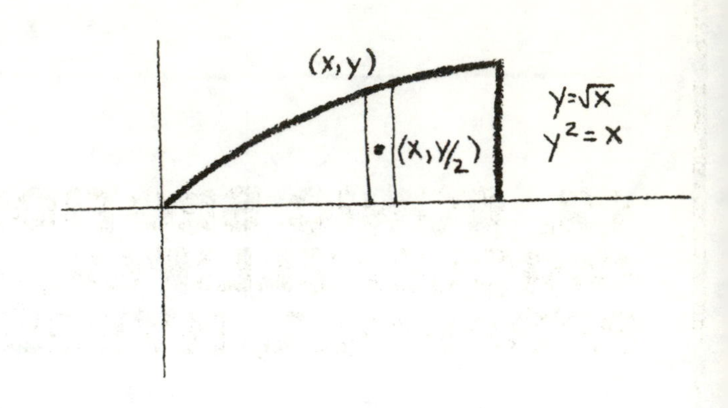

$$\int_0^9 A(x)\,dx \qquad A = \frac{1}{2}\pi r^2 = \frac{1}{2}\pi(y/2)^2 = \frac{\pi}{8}y^2 = \frac{\pi x}{8}$$

$$\frac{\pi}{8}\int_0^9 x\,dx = \frac{\pi x^2}{16}\Big[_0^9 = \frac{81\pi}{16}$$

Since we are adding circular slices (well to be absolutely accurate, semicircular slices), the volume does have a pi in it.

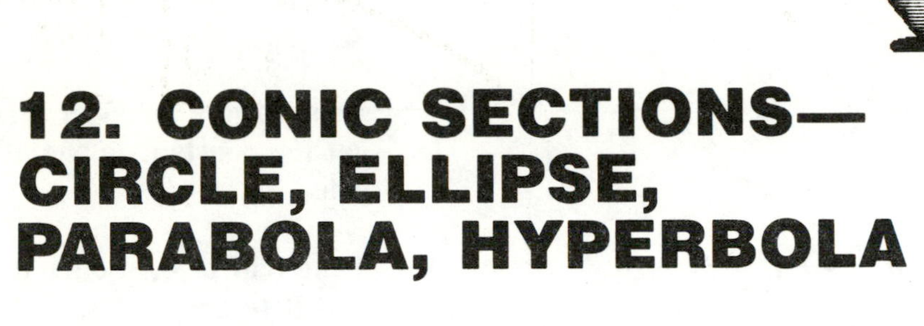

12. CONIC SECTIONS—CIRCLE, ELLIPSE, PARABOLA, HYPERBOLA

Most books call the circle, parabola, ellipse, and hyperbola CONIC SECTIONS without explaining why. These curves are found by passing a plane through the truncated (cut off) right circular cone pictured here. They are formed as follows:

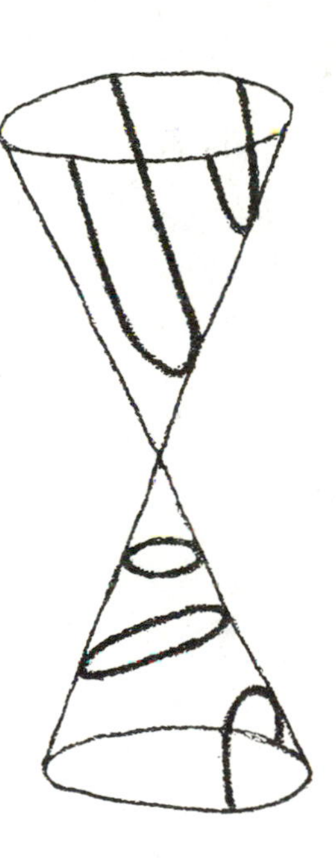

Circle — plane parallel to the top or bottom

Ellipse — plane on the top or bottom not parallel to the top or bottom but hitting all parts of the outside

Parabola — a plane parallel to an edge of the cone

Hyperbola — formed by a plane intersecting the top and bottom

Definition CIRCLE — The set of points that are equidistant from a point, the center (h,k). That distance is r, the radius.

The distance formula:

$$r = [(x-h)^2 + (y-k)^2]^{1/2}$$

Squaring we get $(x-h)^2 + (y-k)^2 = r^2$.

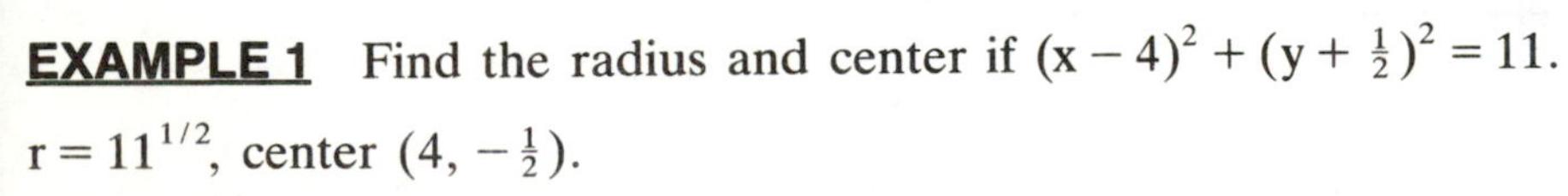

EXAMPLE 1 Find the radius and center if $(x-4)^2 + (y+\frac{1}{2})^2 = 11$.

$r = 11^{1/2}$, center $(4, -\frac{1}{2})$.

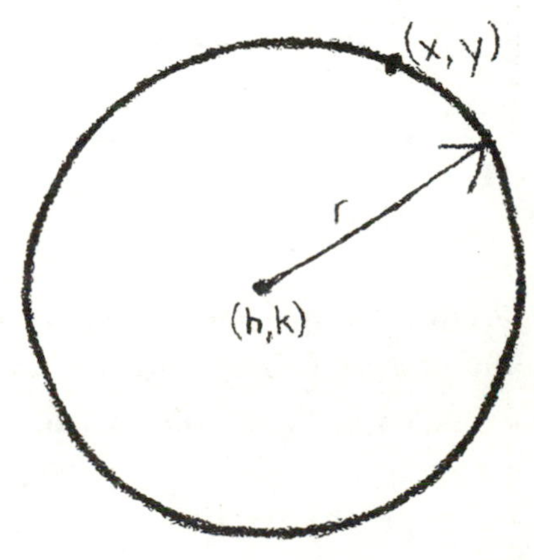

Find the center and radius of the circle $2x^2 + 2y^2 + 8x - 16y + 6 = 0$. In order to do this we have to complete the square, something we have not done since the derivation of the quadratic formula.

EXAMPLE 2

$$2x^2 + 2y^2 + 8x - 16y + 6 = 0$$

$$x^2 + y^2 + 4x - 8y + 3 = 0$$

Divide by the coefficient of x^2

$$x^2 + 4x + y^2 - 8y = -3$$

Group the x terms, y terms, get the constant to the other side

$$x^2 + 4x + 4 + y^2 - 8y + 16 = -3 + 4 + 16$$

Take half of 4, square it, add it to both sides, and take half of −8, square it, add it to both sides

$$(x + 2)^2 + (y - 4)^2 = 17$$

Factor into perfect squares (that was the idea) and add the terms on the right

The center is $(-2,4)$, $r = 17^{1/2}$.

For the parabola, ellipse, and hyperbola, it is essential to relate the equation to the picture. If you do, these curves are very simple.

Definition PARABOLA — the set of all points that are equidistant from a point, called a FOCUS, and a line called a DIRECTRIX.

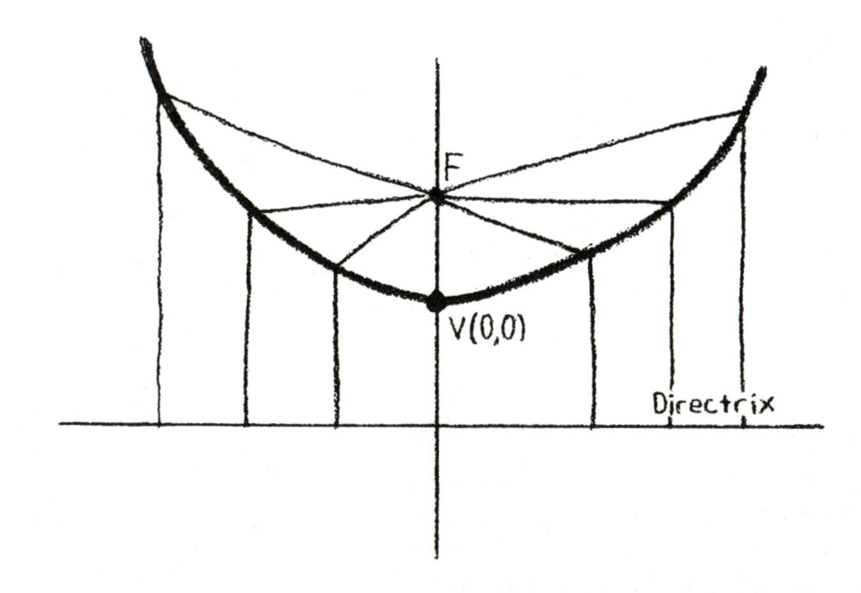

Point V is the vertex, equidistant from the focus and directrix and the closest to the directrix and to the focus.

Let us do this development algebraically. Let the vertex be at $(0,0)$. The focus is $(0,c)$. The directrix is $y = -c$. Let (x,y) be any point on the parabola. The definition of a parabola says $FP = PQ$.

Just like before, everything on PQ has the same x value, and everything on RQ has the same y value. The co-ordinates of Q are $(x,-c)$. Since the x values are the same, the length of $PQ = y-(-c)$. Using the distance formula to get FP and setting it equal to FP, we get $((x-0)^2+(y-c)^2)^{1/2} = y+c$. Squaring we get $x^2+y^2-2cy+y^2 = x^2+2cy+c^2$. Simplifying we get $x^2 = 4cy$.

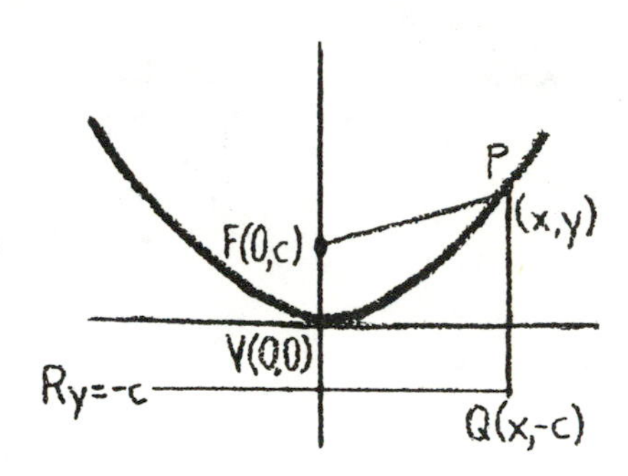

We will make a small chart relating the vertex, focus, directrix, equation, and picture.

	Vertex	Focus	Directrix	Equation	Picture	Comment
1.	(0,0)	(0,c)	$y=-c$	$x^2=4cy$		The original derivation
2.	(0,0)	(0,−c)	$y=c$	$x^2=-4cy$		y replaced by −y
3.	(0,0)	(c,0)	$x=-c$	$y^2=4cx$		x,y interchange in 1
4.	(0,0)	(−c,0)	$x=c$	$y^2=-4cx$		x replaced by −x in 3

If you relate the picture to the original equation, the sketching will be easy.

EXAMPLE 3 Given $y^2=-7x$. Sketch. Label vertex, focus, directrix.

From the chart, we know the sketch is picture 4. Now let $4c=7$ (ignore the minus sign). $c=7/4$. The vertex is (0,0). The focus is $(-7/4,0)$, because it is on x-axis to the left of the origin. The directrix is $y=7/4$; y, a vertical line, $= +7/4$, because it is to the right of the origin.

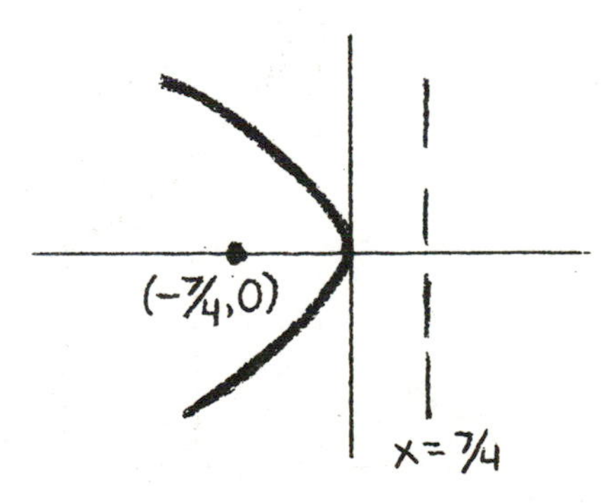

EXAMPLE 4 Sketch $(y-3)^2=-7(x+2)$.

To understand the following we need only note the difference between $x^2+y^2=25$ and $(x-3)^2+(y+6)^2=25$. Has the shape changed? No. Has the radius changed? No. What has changed? The center. Instead of being at the point (0,0), the center is at the point (3,−6).

In the case of the parabola, what has changed is the vertex. Instead of being at the point (0,0), the vertex is at the point (−2,3). The shape is the same. 4c still = 7. So $c=7/4$. The focus now becomes $(-2-7/4,3)$, 7/4 to the left of the vertex (−7/4 from the x co-ordinate). The directrix is $x=-2+7/4$.

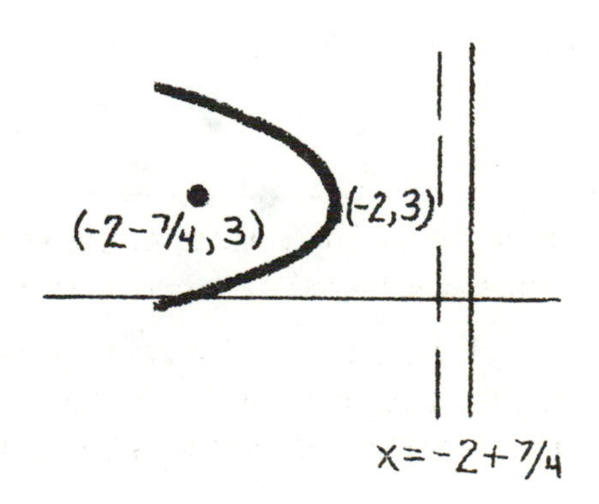

EXAMPLE 5 Sketch the parabola $2x^2 + 8x + 10 = 0$.

$2x^2 + 8x + 6y + 10 = 0$	**Original**
$x^2 + 4x + 3y + 5 = 0$	**Divide through by the coefficient of the squared variable**
$x^2 + 4x = -3y - 5$	**On one side, get all the terms that have the squared letter; everything else to the other side**
$x^2 + 4x + 4 = -3y - 5 + 4$	**Complete the square; add to both sides**
$(x+2)^2 = -3y - 1$	**Factor and simplify**
$(x+2)^2 = -3(y + 1/3)$	**Weird thing. No matter what the coefficient on the right side, factor the whole coefficient out, even if there is a fraction in the parentheses**

Sketch $v(-2,-1/3)$, shape 2, ∩. $4c = 3$, $c = 3/4$. $F(-2,-1/3,-3/4)$. Directrix $y = -1/3 + 3/4$.

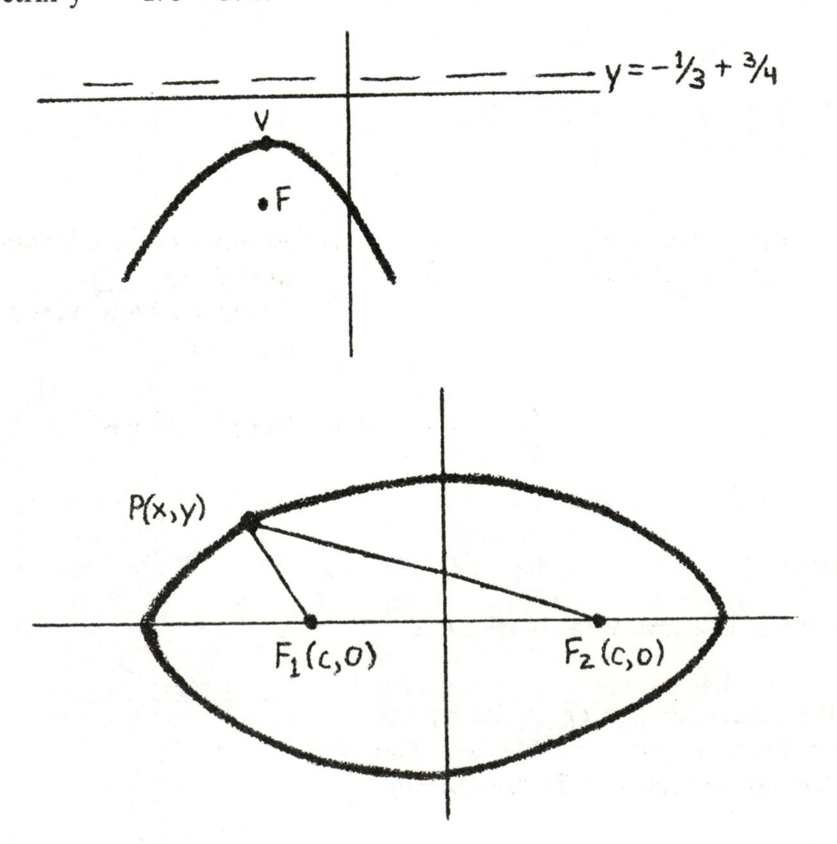

We will now look at the ellipse. Algebraically the ELLIPSE is defined as $PF_1 + PF_2 = 2a$, where $2a > 2c$, the distance between F_1 and F_2. In words, given 2 points, F_1 and F_2, 2 foci. If we find all points P, such that if we go from F_1 to P and then from P to F_2, add those 2 distances together, and always get the same number, 2a, where a will be determined later, we will get an ellipse.

I know you would desperately like to know how to draw an ellipse. This is how. Take a non-elastic string. Attach both ends with thumbtacks to the table. Take the pencil point and stretch the string as far as it will go. Go 360 degrees. You will trace out an ellipse.

Some of you have seen the equation for an ellipse, but few of you have seen the derivation. It is an excellent algebraic exercise for you to try. You will see there is a lot that goes into a rather simple equation.

$$PF_1 + PF_2 = 2a$$

$$\sqrt{[x-(-c)]^2+(y-0)^2} + \sqrt{(x-c)^2+(y-0)^2} = 2a$$

$$[\sqrt{(x+c)^2+y^2}]^2 = [2a - \sqrt{(x-c)^2+y^2}]^2$$

$$x^2 + 2cx + c^2 + y^2 = 4a^2 + x^2 - 2cx + c^2 + y^2 - 4a\sqrt{(x-c)^2+y^2}$$

$$4cx - 4a^2 = -4a\sqrt{(x-c)^2+y^2}$$ **Combine like terms; isolate the radical**

$$(cx - a^2)^2 = [-a\sqrt{(x-c)^2+y^2}]^2$$

$$a^4 - 2a^2cx + c^2x^2 = a^2(x^2 - 2cx + c^2 + y^2)$$

$$a^4 - a^2c^2 = a^2x^2 - c^2x^2 + a^2y^2$$

$$\frac{(a^2-c^2)x^2}{(a^2-c^2)a^2} + \frac{a^2y^2}{a^2(a^2-c^2)} = \frac{a^2(a^2-c^2)}{a^2(a^2-c^2)}$$ **Reverse sides; take out common factors**
Divide on both sides by $a^2(a^2 - c^2)$

$$\frac{x^2}{a^2} + \frac{y^2}{a^2-c^2} = 1$$ **Let $a^2 - c^2 = b^2$**

Finally we get $\frac{x^2}{a^2} + \frac{y^2}{b^2} = 1$. Whew!

We are still not finished. What is a and what is b? Let's investigate.

Since T is any point on the ellipse, $F_1T + TF_2 = 2a$. By symmetry, $F_1T = TF_2$. So $F_1T = a$. Since $a^2 - c^2 = b^2$, $GT = b$. The coordinates of T are (0,b), and the co-ordinates of T′ are (0,−b).

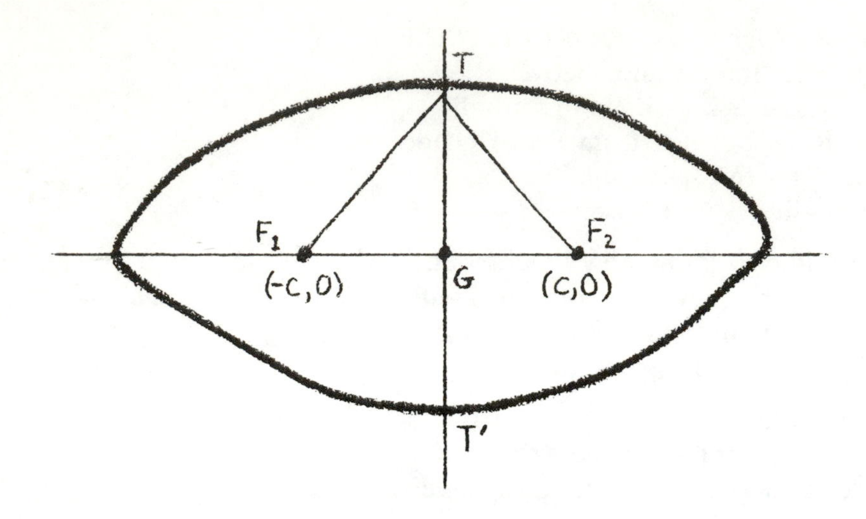

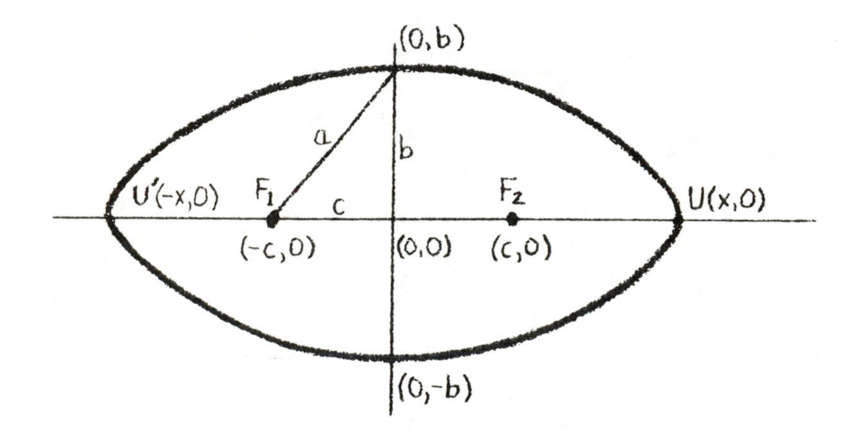

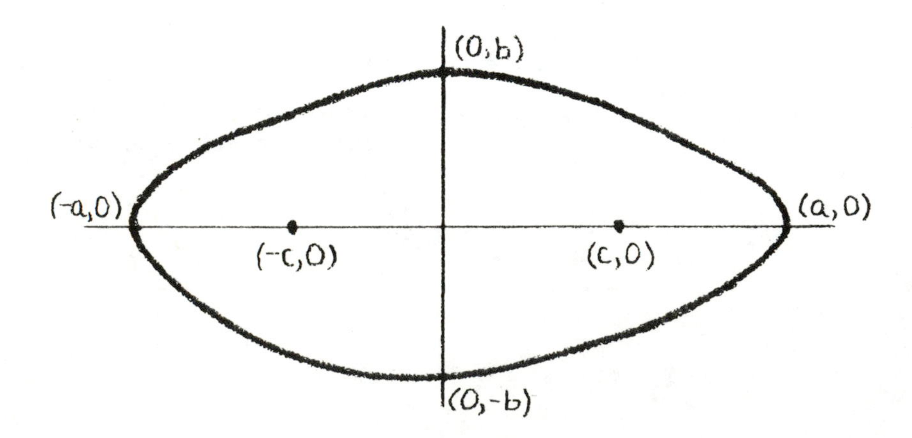

We would like to find the coordinates of U and we have used up the letters a,b,c. Oh well, let's see what happens. $F_2U + UF_1 = 2a$. $F_2U = x - c$. $UF_1 = x + c$. $x + c + x - c = 2a$. So $x = a$. The coordinates of U are (a,0). The coordinates of U′ are (−a,0).

c = half the distance between the foci. b = length of SEMIMINOR axis, ("semi" means half, "minor" means smaller, "axis" means line). a = length of the SEMIMAJOR axis = distance from a focus to a minor vertex. $(\pm a,0)$ are the MAJOR VERTICES. $(0,\pm b)$ are the MINOR VERTICES or CO-VERTICES. $(\pm c,0)$ are the foci.

Although the derivation is very long, sketching should be short.

EXAMPLE 6 Sketch $x^2/11 + y^2/8 = 1$.

In the case of an ellipse, the longer axis is indicated by which number is larger under x^2 or y^2. That term is a^2. (Try not to remember a or b—remember the picture.) This ellipse is longer in the x direction.

Letting $y = 0$, we get rhe major vertices $(\pm\sqrt{11},0)$. Letting $x = 0$, we get the minor vertices $(0,\pm\sqrt{8})$. $c = \sqrt{11-8}$. The foci are $(\pm\sqrt{3},0)$. The sketch is below.

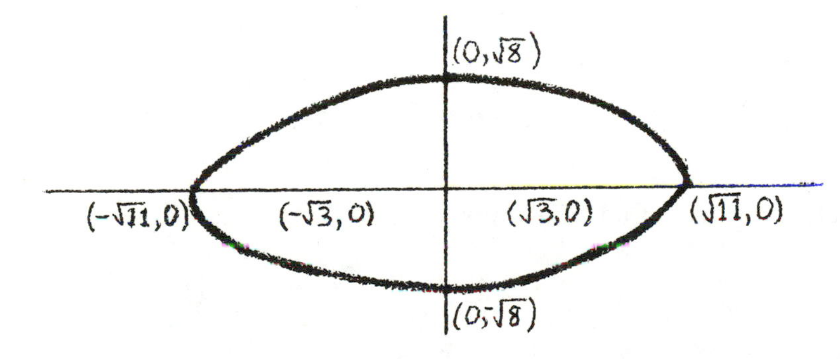

EXAMPLE 7 Sketch $x^2/5 + y^2/26 = 1$.

Major vertices $(0,\pm\sqrt{26})$. Minor vertices $(\pm\sqrt{5},0)$. Foci $(0,\pm\sqrt{21})$—foci are always on the longer axis.

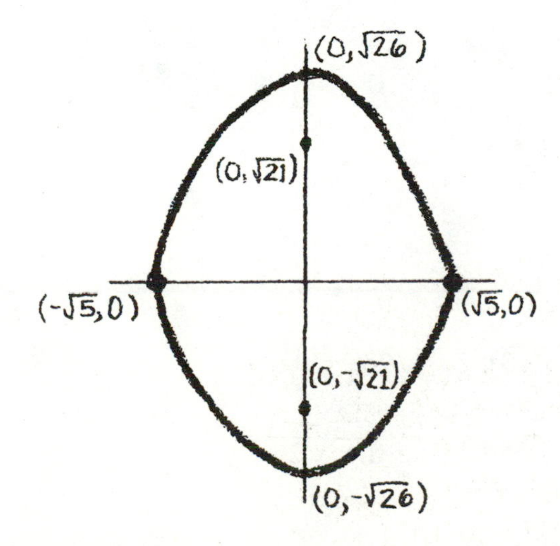

EXAMPLE 8 $\frac{(x-6)^2}{11} + \frac{(y+4)^2}{8} = 1$

This is the same as Example 12 except the center is no longer at the point (0,0). It is moved to the point (6,−4). Now the major vertices are $(6 \pm \sqrt{11}, -4)$. The minor vertices are $(6, -4 \pm \sqrt{8})$. The foci are $(6 \pm \sqrt{3}, -4)$.

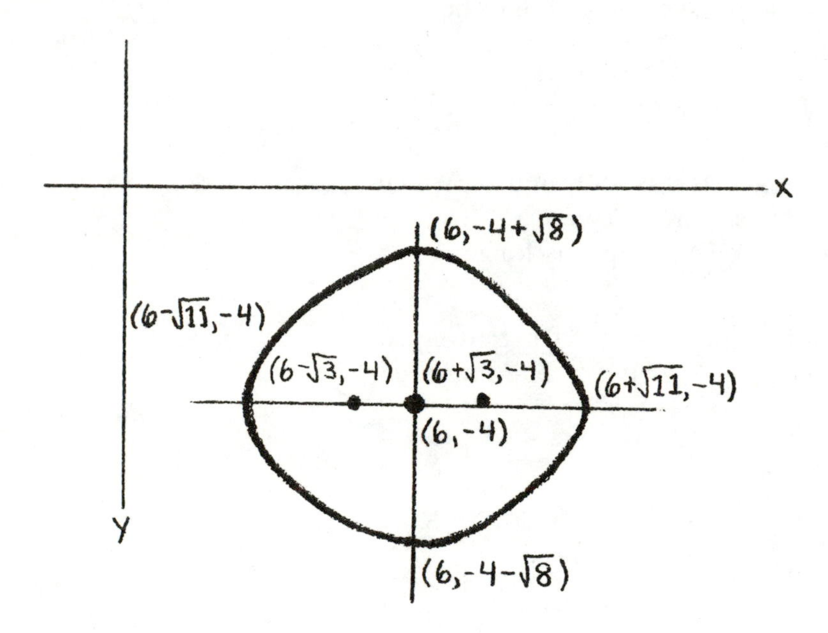

Weird numbers are intentionally chosen so that you know exactly where the numbers come from.

EXAMPLE 9 Sketch and discuss $4x^2 + 5y^2 + 30y - 40x + 45 = 0$.

Here we must complete the square in a slightly different manner.

$$4x^2 + 5y^2 + 30y - 40x + 45 = 0$$

$$4x^2 - 40x + 5y^2 + 30y = -45$$

$$4[x^2 - 10x + (-10/2)^2] + 5[y^2 + 6y + (6/2)^2]$$

$$= -45 + 4(-10/2)^2 + 5(6/2)^2$$

$$\frac{4(x-5)^2}{100} + \frac{5(y+3)^2}{100} = \frac{100}{100}$$

$$\frac{(x-5)^2}{25} + \frac{(y+3)^2}{20} = 1$$

Center (5,−3). Vertices $(5 \pm \sqrt{25}, -3)$, $(5, -3 \pm \sqrt{20})$. $c = \sqrt{25-20} = \sqrt{5}$. Foci $(5 \pm \sqrt{5}, -3)$.

Of course you should put 5 instead of $\sqrt{25}$. I leave the $\sqrt{25}$ so you know where it came from.

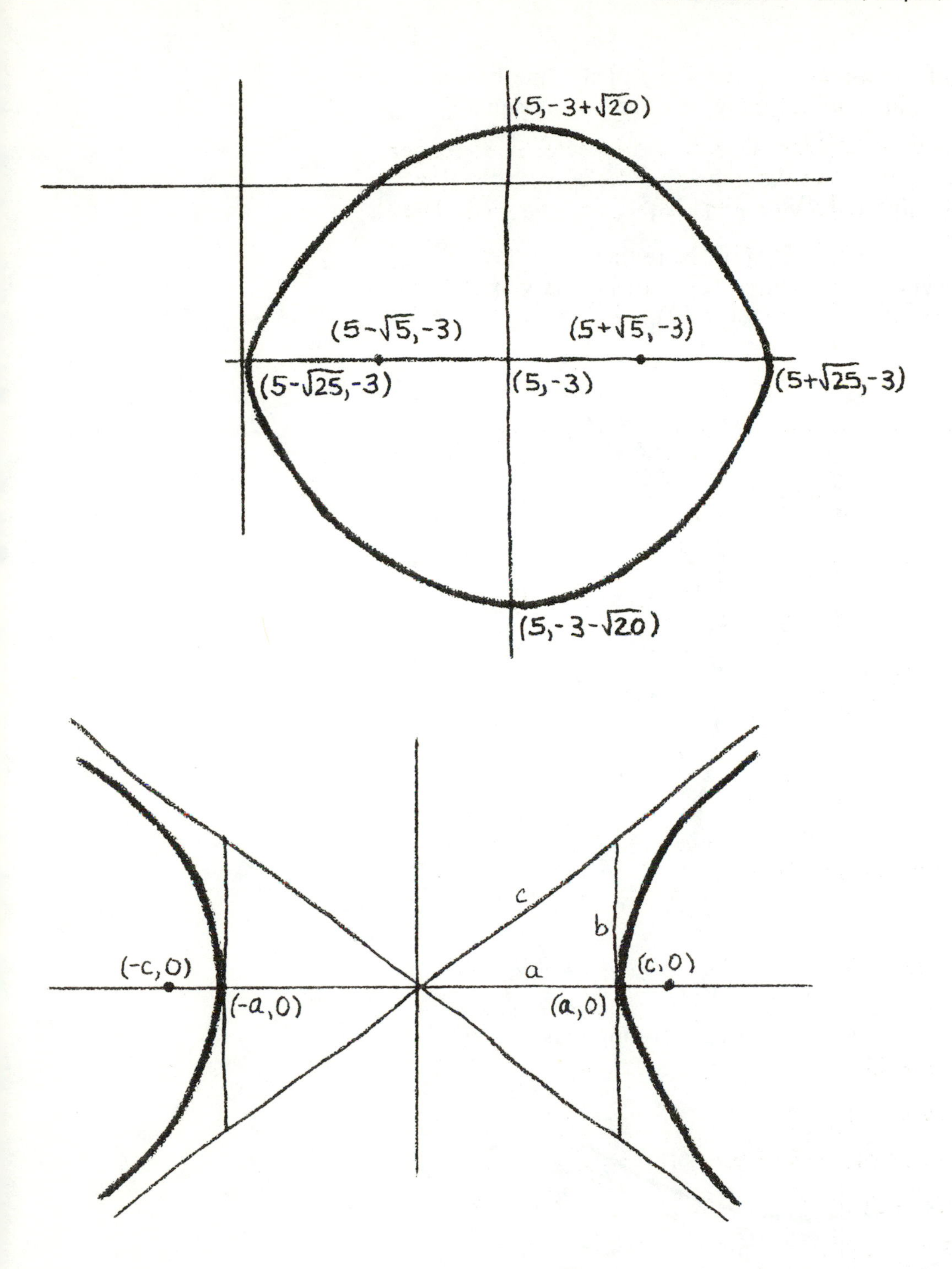

The definition of the HYPERBOLA is $F_1P - PF_2 = 2a$ where the foci are $(\pm c,0)$. The derivation is exactly the same as for an ellipse. Once is enough!! The equation we get is $x^2/a^2 - y^2/b^2 = 1$, where $a^2 + b^2 = c^2$. $(\pm a,0)$ are called the TRANSVERSE VERTICES. The hyperbola has asymptotes $y = \pm(b/a)x$.

Note 1: The shape of a hyperbola is determined by the location of the minus sign, not which number is larger under the x^2 or y^2.

Note 2: In the case of the asymptote, the slope of the line b/a is the square root of the number under the y^2 divided by the square root of the term under the x^2 term.

EXAMPLE 10 Sketch and label $x^2/7 - y^2/11 = 1$.

Transverse vertices $y = 0$, $x = \pm\sqrt{7}(\pm\sqrt{7},0)$. Note that if $x = 0$, $y = \pm\sqrt{-11}$, which are imaginary. The curve does not hit the y-axis. $c = \sqrt{7+11}$ foci are $(\pm\sqrt{18},0)$. $y = \pm(\sqrt{11}/\sqrt{7}),x$.

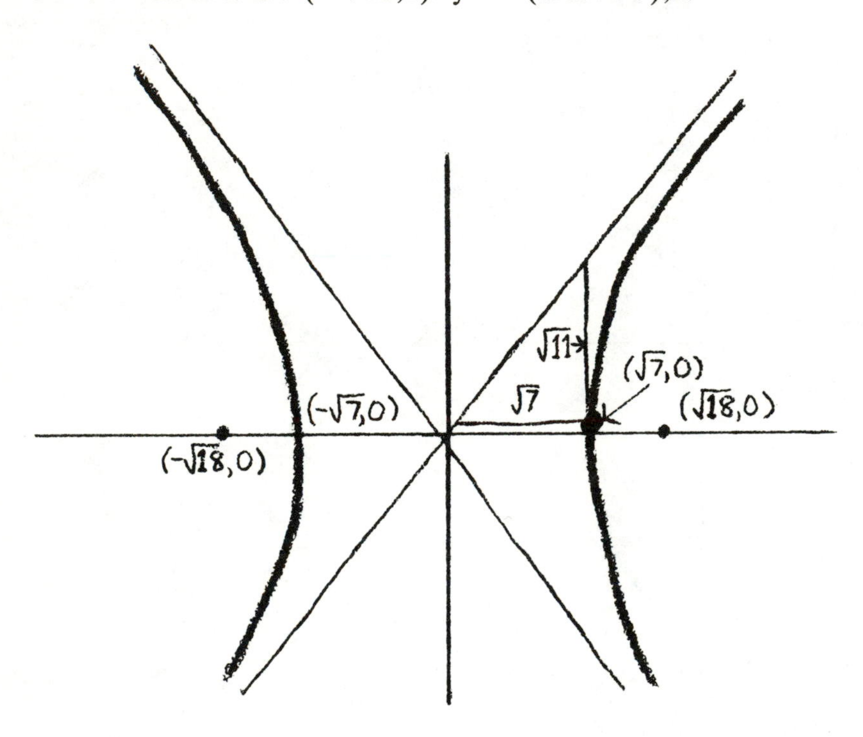

EXAMPLE 11 Sketch and discuss $y^2/5 - x^2/17 = 1$.

Transverse vertices $(0,\pm\sqrt{5})$. Foci $(0,\pm\sqrt{22})$. Asymptotes $y = \pm(\sqrt{5}/\sqrt{17})x$.

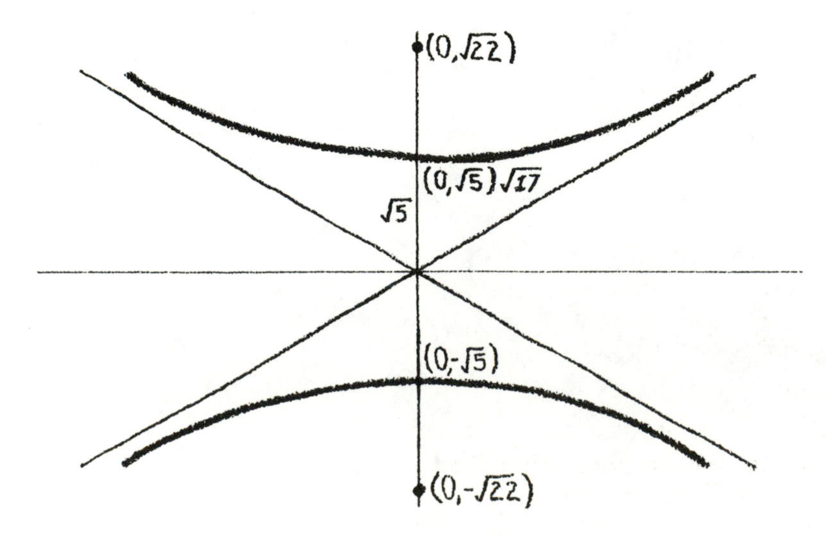

EXAMPLE 12 The same form $\frac{(y-6)^2}{5} - \frac{(x+7)^2}{17} = 1$.

This is the same as the previous sketch except the "center of the hyperbola", the place where the asymptotes cross, is no longer (0,0). It is now $(-7,6)$. Transverse vertices $(-7,6 \pm \sqrt{5})$, foci $(-7,6 \pm \sqrt{22})$, asymptotes $(y-6) = \pm(\sqrt{5}/\sqrt{17})(x+7)$.

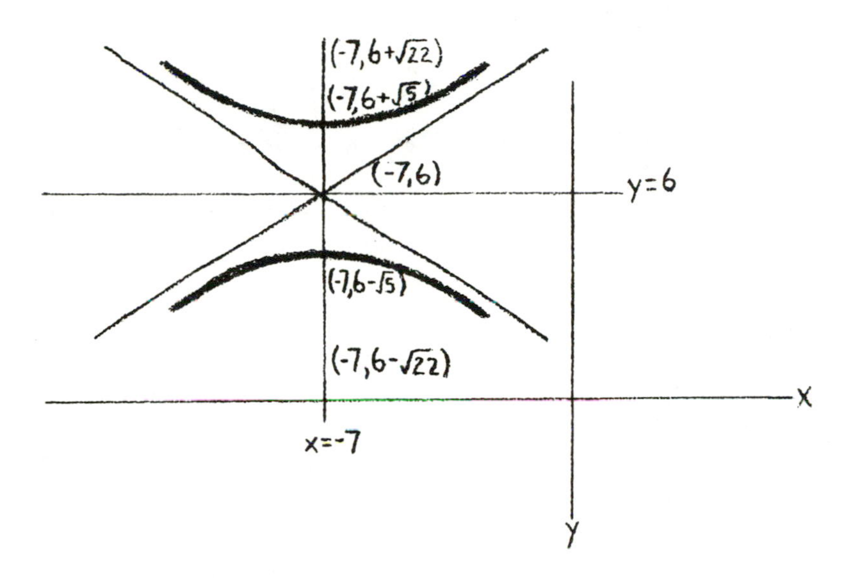

EXAMPLE 13 Sketch and discuss $25x^2 - 4y^2 + 50x - 12y + 116 = 0$.

For the last time we will complete the square, again a little bit differently.

$$25x^2 - 4y^2 + 50x - 12y + 116 = 0$$

$$25x^2 + 50x - 4y^2 - 12y = -116$$

$$25[x^2 + 2x + (2/2)^2] - 4[y^2 + 3y + (3/2)^2]$$

$$= -116 + 25(2/2)^2 - 4(3/2)^2$$

$$\frac{25(x+1)^2}{-100} - \frac{4(y+3/2)^2}{-100} = \frac{-100}{-100}$$

$$\frac{(y+3/2)^2}{25} - \frac{(x+1)^2}{4} = 1$$

Center $(-1,-3/2)$, vertices $(-1,-3/2 \pm \sqrt{25})$, foci $(-1,-3/2 \pm \sqrt{29})$. Asymptotes $y + 3/2 = \pm(\sqrt{25}/\sqrt{4})(x+1)$.

Sometimes we have a puzzle. Given some information can we find the equation? You must always draw the picture and relate the picture to its equation.

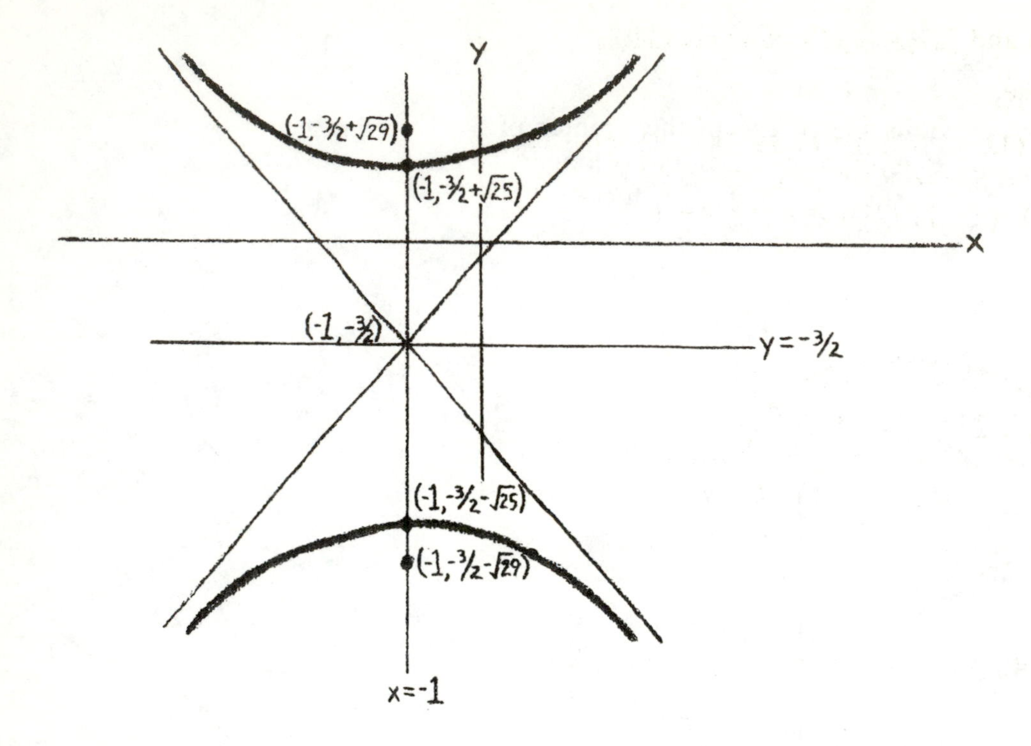

EXAMPLE 14 Find the equation of the parabola with focus (1,3), directrix $x = 11$.

Drawing F and the directrix, the picture must be the following picture on the bottom. Vertex is halfway between the x numbers. So $x = (11 + 1)/2 = 6$. V (6,3). c = the distance between V and F = 5. The equation is $(y - 3)^2 = -4c(x - 6) = -20(x - 6)$. Remember the minus sign is from the shape and c is always positive for these problems.

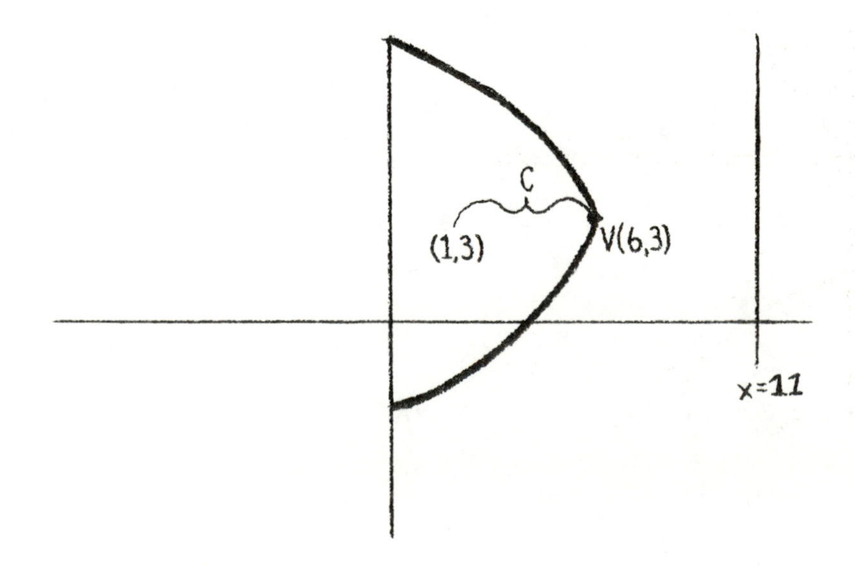

EXAMPLE 15 Vertices (2,3) and (12,3) and one focus (11,3).

Find the equation of the ellipse.

Two vertices give the center, $(12+2)/2,3) = (7,3)$. F(11,3). $(x-7)^2/a^2 + (y-3)^2/b^2 = 1$. $a = 12 - 7 = 5$. $c = 11 - 7 = 4$. $a^2 - b^2 = c^2$. $5^2 - b^2 = 4^2$. $b^2 = 9$ (no need for b). $(x-7)^2/25 + (y-3)^2/9 = 1$.

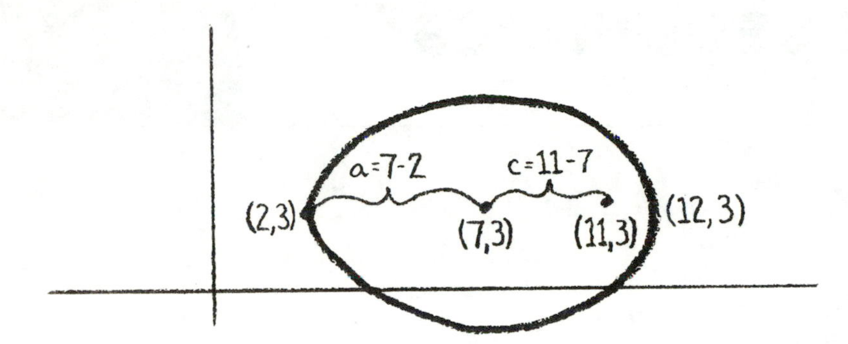

EXAMPLE 16 Find the equation of the hyperbola with vertices $(0,\pm6)$ and asymptotes $y = \pm(3/2)x$.

$V(0,\pm6)$ says the center is (0,0) and its shape is $y^2/36 - x^2/a^2 = 1$. The slope of the asymptotes is $3/2$ = square root of the number under y^2 over the square root of the number under x^2 term. So $3/2 = 6/?$. So $? = 4$. So $a^2 = 4^2 = 16$. The equation is $y^2/36 - x^2/16 = 1$.

This kind of question is shorter in length, but it does take practice. So practice!!!

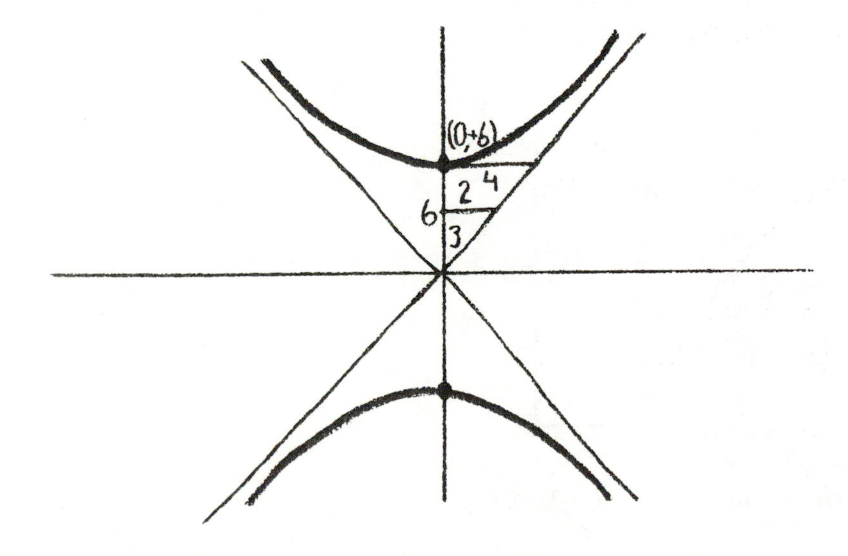

13. INFINITE SEQUENCES

This topic brings some controversy. Some people think it is very difficult; some, very easy. I believe if you understand the beginning, the rest of the chapter is not too bad.

Definition SEQUENCE — A sequence of terms, technically, is a function for which the domain is the positive integers. Non-technically, there is a first term called a_1 (read "a sub one," where the "one" is a subscript, not an exponent, denoting the first term), a_2 ("a sub two") denoting the second term, and so on. The notation for an infinite sequence is $\{a_n\}$.

Let us give some examples. We will list some sequences, write the first 4 terms, and then term number 100 by substituting 1,2,3,4, . . . ,100 for n in a_n.

EXAMPLE 1

$\{a_n\}$	1st	2nd	3rd	4th	100th
$\left\{\frac{n}{n+1}\right\}$	$\frac{1}{2}$	$\frac{2}{3}$	$\frac{3}{4}$	$\frac{4}{5}$	$\frac{100}{101}$
$\left\{\frac{(-1)^{n+1}(4n+1)}{n^2+1}\right\}$	$\frac{5}{2}$	$\frac{-9}{5}$	$\frac{13}{10}$	$\frac{-17}{17}$	$\frac{-401}{10001}$
$\{6\}$	6	6	6	6	6

Definition (nontechnical) We write $\lim_{n\to\infty} a_n = L$ if the larger n gets, the closer a_n gets to L.

In this case we say that $\{a_n\}$ converges to L (or has the limit L). If a_n goes to plus or minus infinity or does not go to a single number, then $\{a_n\}$ diverges (or has NO limit).

EXAMPLE 2 Find the limit of $\{(n+9)/n^2\}$.

$a_n = (n+9)/n^2 = 1/n + 9/n^2$. As n goes to infinity, both terms go to 0. Therefore the sequence converges to 0.

EXAMPLE 3 Find the limit of $\{(2n^2+3n+2)/(5-7n^2)\}$.

Divide top and bottom of a_n by n^2. We get $[2+(3/n)+(2/n^2)]/[(5/n^2-7)]$. As n goes to infinity, a_n goes to $2/(-7)$. The sequence has the limit $-2/7$.

Note: This should look very familiar. This is how we found horizontal asymptotes. Also note that we can use L'Hopital's rule.

EXAMPLE 4 Find the limit of $\{a_n\} = \{\ln(n+1) - \ln n\}$.

$$\lim_{n\to\infty} \ln(n+1) - \ln n = \lim_{n\to\infty} \ln[(n+1)/n]$$
$$= \lim_{n\to\infty} \ln(1+1/n) = \ln 1 = 0$$

The sequence converges to 0.

EXAMPLE 5 Does $\{(-1)^n\}$ have a limit?

This sequence is $-1,+1,-1,+1,-1,\ldots$. There is no limit because the sequence does not go to 1 number.

Definition (technical) $\lim_{n\to\infty} a_n = L$ if given an $\varepsilon > 0$, there exists an $N > 0$, such that if $n > N$, $|a_n - L| < \varepsilon$.

Note: It is not important that you know the technical definition of a limit to understand the rest of the chapter. But . . . at this point of your mathematical career, you should start understanding the background. It probably will help you later on. It would also be nice if you could see the beauty and the depth of this material—the beginnings of calculus. It truly is a wonderful discovery.

EXAMPLE 6 Using ε,N, show $\lim_{n\to\infty} (2n+5)/(n+1) = 2$.

$$\left|\frac{2n+5}{n+1} - 2\right| = \left|\frac{2n+5}{n+1} - \frac{2(n+1)}{n+1}\right| = \left|\frac{3}{n+1}\right| < \varepsilon$$

provided $3/\varepsilon < n+1$ or $3/\varepsilon - 1 < n$. We then choose N as the whole-number part of $3/\varepsilon - 1$.

The following theorems are used often. They are proved in many books and will only be stated here.

Theorems: Let $\lim_{n\to\infty} a_n = L$, $\lim_{n\to\infty} b_n = M$. $k =$ constant, f continuous. Then

1. $\lim_{n\to\infty} (a_n \pm b_n) = L \pm M$
2. $\lim_{n\to\infty} (a_n b_n) = LM$
3. $\lim_{n\to\infty} (a_n/b_n) = L/M \qquad M \neq 0$
4. $\lim_{n\to\infty} ka_n = kL$
5. $\lim_{n\to\infty} f(a_n) = f(L)$
6. $c_n \leq d_n \leq e_n$ and $\lim_{n\to\infty} c_n = \lim_{n\to\infty} e_n = P$. Then $\lim_{n\to\infty} d_n = P$.

EXAMPLE 7 Show $\lim_{n\to\infty} (\sin n)/n = 0$.

Using part 6 above, $-1 \leq \sin n \leq 1$. So $-1/n \leq (\sin n)/n \leq 1/n$. As n goes to infinity, $-1/n$, $1/n$ go to 0. Therefore so does $(\sin n)/n$.

Definition 1 An INCREASING SEQUENCE is one where $a_n < a_{n+1}$ for all n.

Definition 2 A NONDECREASING SEQUENCE is one where $a_n \leq a_{n+1}$ for all n.

Similarly we can define DECREASING and NONINCREASING.

Definition 3 A sequence is bounded if $|a_n| \leq M$, some number M and all n.

Another theorem: Every bounded increasing (decreasing) sequence has a limit.

INFINITE SERIES

I know this is getting to be a drag, but it is essential to understand the terminology. This understanding will make the rest of the chapter MUCH easier. I don't know why, but it really seems to.

Definition PARTIAL SUMS — Given sequence $\{a_n\}$.

1st partial sum $S_1 = a_1$

2nd partial sum $S_2 = a_1 + a_2$

3rd partial sum $S_3 = a_1 + a_2 + a_3$

nth partial sum $S_n = a_1 + a_2 + a_3 + \cdots + a_n = \sum_{k=1}^{n} a_k$

The *infinite series* $a_1 + a_2 + a_3 + \cdots$ or $\sum_{k=1}^{\infty} a_k$ is said to *converge to the sum* S if $\lim_{n\to\infty} S_n = S$. If S does not exist, the series *diverges*.

EXAMPLE 8 .767676. . . .

We can write this as an infinite series. $.76 + .0076 + .000076 + \cdots$. This is a geometric series (infinite). This is one of the few series we can find the exact sum of.

$S = a/(1-r) \qquad a = .76 \qquad r = .01 \qquad S = .76/(1-.01) = 76/99$

More generally, the series $a + ar + ar^2 + ar^3 + \cdots$ converges to $a/(1-r)$ if $|r| < 1$.

EXAMPLE 9 $4 - 8 + 16 - 32 + \cdots$: $a = 4$, $r = -2$; diverges.

EXAMPLE 10 $1 + 1 + 1 + 1 + 1 + \cdots$: $a = 1$, $r = 1$; diverges.

EXAMPLE 11 $1 - 1 + 1 - 1 + 1 - \cdots$: $a = 1$, $r = -1$; diverges.

Note that test 1 (below) implies divergence in these 3 examples.

EXAMPLE 12 $\sum_{k=1}^{\infty} \frac{1}{k(k+1)}$

Using partial fractions

$$\frac{1}{k(k+1)} = \frac{1}{k} - \frac{1}{k+1}$$

Writing out the first few terms plus the n − 1 term plus the nth term, we get

$$S_n = (1/1 - 1/2) + (1/2 - 1/3) + (1/3 - 1/4) + \cdots + [1/n - 1/(n+1)]$$

Notice all the middle terms cancel out in pairs. So only the first and last terms remain:

$S_n = 1 - 1/(n+1) \qquad S = \lim_{n\to\infty} S_n = 1$

Again this is one of the few sequences we can find the exact value for. (This is called a telescoping series because it collapses like one of those toy or portable telescopes.) From this point on for almost all of the converging series, we will be able to tell the series converges, but not find its value. Later we will do some approximating.

EXAMPLE 13 $\sum_{k=1}^{\infty} \frac{4(2^k) + 5^k}{7^k}$

After splitting we get 2 geometric series:

$$S = \frac{4(2/7)}{1 - 2/7} + \frac{5/7}{1 - 5/7} = 8/5 + 5/2 = 41/10$$

Theorem: If $\sum_{n=1}^{\infty} a_n = L$ and $\sum_{n=1}^{\infty} b_n = M$, then $\sum_{n=1}^{\infty} (ca_n + b_n) = cL + M$.

Now that we have an idea about what a sequence is and what an infinite series is (hopefully a *very good* idea), we would like to have some tests for when a series converges or diverges.

Test 1. It is necessary that $a_k \to 0$ for $\sum_{k=1}^{\infty} a_k$ to converge.

Note 1: If a_k does not go to 0, $\sum_{k=1}^{\infty} a_k$ diverges.

Note 2: If a_k does go to 0, and that is all we know, we know nothing.

EXAMPLE 14 Tell whether $\sum_{k=1}^{\infty} k/(k+1)$ converges.

$k/(k+1)$ goes to 1. Therefore $\sum_{k=1}^{\infty} k/(k+1)$ diverges.

EXAMPLE 15 The harmonic series $\sum_{k=1}^{\infty} 1/k$

Since $1/k$ goes to 0, we don't know if this series converges or diverges. We shall shortly show the harmonic series diverges.

EXAMPLE 16 The p_2 series $\sum_{k=1}^{\infty} 1/k^2$

Since $1/k^2$ goes to 0, again we can't tell. Shortly we shall show the p_2 series converges.

Test 2. Given a_k, $a_k > 0$, a_k goes to 0 for k big enough. Suppose we have a continuous function f(x) such that $f(k) = a_k$. Then $\sum_{k=1}^{\infty} a_k$ and $\int_1^{\infty} f(x)\,dx$ either both converge or both diverge.

This theorem is easily explained by examples.

EXAMPLE 17 Tell whether $\sum_{k=1}^{\infty} ke^{-k^2}$ converges or diverges.

The improper integral associated with $\sum_{k=1}^{\infty} ke^{-k^2}$ is $\int_1^{\infty} xe^{-x^2}\,dx$. Letting $u = -x^2$, $du = -2x\,dx$.

$$\int_1^{\infty} xe^{-x^2}\,dx = \lim_{b\to\infty} \int_1^b \frac{-2xe^{-x^2}}{-2}\,dx = \lim_{b\to\infty} (-\tfrac{1}{2}) \int_{-1}^{-b^2} e^u\,du$$

$$= \lim_{b\to\infty} -\tfrac{1}{2}(e^{-b^2} - e^{-1}) = \frac{1}{2e}$$

Since the improper integral converges, so does the infinite series.

Note: The value of the improper integral is not the value of the infinite series. But we can say the following: If the integral and the series together converge, then $\int_1^{\infty} f(x)\,dx \le \sum_{k=1}^{\infty} a_k \le a_1 + \int_1^{\infty} f(x)\,dx$.

EXAMPLE 17 continued The bounds on $\sum_{k=1}^{\infty} ke^{-k^2}$ are $\frac{1}{2e} \le \sum_{k=1}^{\infty} ke^{-k^2} \le \frac{1}{e} + \frac{1}{2e}$.

Note: In this case, this is not too good an approximation.

We will get a better one if we take the 4th partial sum, $= \frac{1}{e} + \frac{2}{e^4} + \frac{3}{e^9} + \frac{4}{e^{16}}$. The "error," the estimate on the rest of the terms, is that

$$\sum_{k=5}^{\infty} ke^{-k^2} \le a_5 + \int_5^{\infty} xe^{-x^2}\,dx = \frac{5}{e^{25}} + \frac{1}{2e^{25}}\ 1.4 \times 10^{-11}!!!!$$

This is more accuracy than you will probably ever need!!! Lots of things you cannot even integrate.

EXAMPLE 18 The harmonic series $\sum_{k=1}^{\infty} 1/k$ diverges.

$\lim_{b\to\infty} \int_1^b 1/x\,dx = \ln b$, which goes to infinity as b goes to infinity.

EXAMPLE 19 The p_2 series $\sum_{k=1}^{\infty} 1/k^2$ converges.

$\int_1^b 1/x^2\,dx = -1/b + 1$. Since $-1/b$ goes to 0 as b goes to infinity, this improper integral converges. So does the p_2.

EXAMPLE 20 $\sum_{k=1}^{\infty} 1/k^p$

If $p > 1$, converges, and if $p \leq 1$, diverges. Just use the integral test. It's easy.

Test 3. The comparison test. Given $\sum_{k=1}^{\infty} a_k$, $\sum_{k=1}^{\infty} b_k$ where $0 < a_k \leq b_k$,

1. If $\sum_{k=1}^{\infty} b_k$ converges, so does $\sum_{k=1}^{\infty} a_k$.

2. If $\sum_{k=1}^{\infty} a_k$ diverges, so does $\sum_{k=1}^{\infty} b_k$.

Let us talk through part 1. The second part can be shown to be equivalent. The partial sums S_n of the $\sum_{k=1}^{\infty} b_k$ series are uniformly bounded because the first N terms are bounded by their maximum and the rest are bounded by $L + \varepsilon$. Therefore the partial sums of the $\sum_{k=1}^{\infty} a_k$ series also are bounded, being respectfully smaller than those of $\sum_{k=1}^{\infty} b_k$. Moreover, since $a_k > 0$, then the partial sums of the a_k form an increasing sequence. Now we have an increasing bounded sequence which has a limit. Therefore $\sum_{k=1}^{\infty} a_k$ converges.

EXAMPLE 21 Examine $\sum_{k=1}^{\infty} 1/(4 + k^4)$.

$1/(4 + k^4) < 1/k^4$. $\sum_{k=1}^{\infty} 1/k^4$ converges by Example 20. Since the given series is termwise smaller than a convergent series, it must converge by the comparison test.

EXAMPLE 22 Examine $\sum_{k=1}^{\infty} (2 + \ln k)/k$.

$\sum_{k=1}^{\infty} (2 + \ln k)/k > \sum_{k=1}^{\infty} 2/k$ = twice a divergent series (the harmonic). Since the given series is termwise larger than a divergent series, the given series must diverge.

Test 4. Limit comparison test. Given $\sum_{k=1}^{\infty} a_k$ and $\sum_{k=1}^{\infty} b_k$, $a_k \geq 0$, $b_k \geq 0$. If $\lim_{k\to\infty} (a_k/b_k) = r$, where r is any positive number, both series converge or both diverge.

EXAMPLE 23 $\sum_{k=1}^{\infty} 3/(5k^4+4)$

Let us compare this series with $\sum_{k=1}^{\infty} 1/k^4$.

$$\lim_{k\to\infty} [3/(5k^4+4) \text{ divided by } 1/k^4]$$

Divide top and bottom by k^4

$$= \lim_{k\to\infty} \frac{3k^4}{5k^4+4}$$

$$= \lim_{k\to\infty} \frac{3}{5+(4/k^4)} = 3/5$$

Since the limit is a positive number, both series "do the same thing." Since $\sum_{k=1}^{\infty} 1/k^4$ converges, so does $\sum_{k=1}^{\infty} 3/(5k^4+4)$.

Test 5. Ratio test. Given $a_k \geq 0$. $\lim_{k\to\infty} (a_{k+1}/a_k) = r$. (1) $r > 1$, diverges. (2) $r < 1$, converges. (3) $r = 1$, ???? (use another test).

EXAMPLE 24 Examine $\sum_{k=1}^{\infty} k^2/5^k$.

$$a_{k+1}/a_k = (k+1)^2/5^{k+1} \text{ divided by } k^2/5^k$$

$$= \frac{(k+1)^2}{5^{k+1}} \times \frac{5^k}{k^2} = \frac{k^2+2k+1}{5k^2}$$

$$\lim_{k\to\infty} \frac{a_{k+1}}{a_k} = \lim_{k\to\infty} \frac{1+\frac{2}{k}+\frac{1}{k^2}}{5} = \frac{1}{5} < 1 \qquad \sum_{k=1}^{\infty} k^2/5^k \text{ converges}$$

EXAMPLE 25 Examine $\sum_{k=1}^{\infty} 7^k/k!$

Note: 6! means 6(5)(4)(3)(2)(1). ***Also note:*** $(k+1)! = (k+1)(k!)$. That is, $10! = 10(9!)$, etc.

Let us again use the ratio test.

$$a_{k+1}/a_k = 7^{k+1}/(k+1)! \text{ divided by } 7^k/k!$$

$$= \frac{7^{k+1}}{(k+1)!} \times \frac{k!}{7^k} = \frac{7}{k+1}$$

$$\lim_{k\to\infty} \frac{7}{k+1} = 0 < 1 \qquad \sum_{k=1}^{\infty} 7^k/k! \text{ converges}$$

EXAMPLE 26 Let's now look at $\sum_{k=1}^{\infty} k^k/k!$

This is a little trickier than most. We again use the ratio test.

$$a_{k+1}/a_k = (k+1)^{k+1}/(k+1)! \text{ divided by } k^k/k!$$

$$= \frac{(k+1)^{k+1}}{(k+1)!} \times \frac{k!}{k^k} = \frac{(k+1)(k+1)^k k!}{(k+1)k!k^k} = \frac{(k+1)^k}{k^k}$$

$$= (1+1/k)^k$$

Since the $\lim_{k\to\infty} [1+(1/k)]^k = e > 1$, $\sum_{k=1}^{\infty} k^k/k!$ diverges.

EXAMPLE(S) 27 In order to show the third part of the previous theorem, you should apply the ratio test to both the harmonic series and the p_2 series. Both give a ratio of 1. The first series diverges and the second converges. So if the ratio is 1, we must indeed use another test.

Test 6. Root test. Given $\sum_{k=1}^{\infty} a_k$. Take $\lim_{k\to\infty} (a_k)^{1/k} = r$. ($a_k \geq 0$.) (1) $r > 1$, diverges. (2) $r < 1$, converges. (3) $r = 1$, ???????

Note: To show (3) we would again use the harmonic and p_2 series. Let us give examples for (1) and (2).

EXAMPLE 28 $\sum_{k=1}^{\infty} 3^k/k^k$

Take $(3^k/k^k)^{1/k} = 3/k$. $\lim_{k\to\infty} (3/k) = 0 < 1$. So the series converges.

EXAMPLE 29 $\sum_{k=1}^{\infty} 2^k/k^2$

Take $(2^k/k^2)^{1/k} = 2/k^{2/k}$. $\lim_{k\to\infty} 2/k^{2/k} = 2/1 = 2$ ($\lim_{k\to\infty} k^{2/k} = \lim_{k\to\infty} e^{2(\ln k)/k} = e^{2(0)} = 1$). Since $2 > 1$, the series diverges.

Up to this time we have dealt exclusively with positive terms. Now we will deal with infinite series which have terms that alternate from positive to negative. We will assume the first term is positive. The notation will be as follows: alternating series $\sum_{k=1}^{\infty} (-1)^{k+1} a_k$, where all a_k are positive.

Test 7. Given an alternating series where (a) $0 < a_{k+1} \leq a_k$, $k =$ 1-2-3-4, ... (b) $\lim_{k\to\infty} a_k = 0$. Then the series converges to S and $S \leq a_1$.

In clearer English, the ONLY thing you must do to show an alternating series converges is to show the terms go to zero. (If only all series were that easy!)

EXAMPLE 30 Alternating harmonic $\sum_{k=1}^{\infty} \frac{(-1)^{k+1}}{k}$ converges since the terms go to 0.

Definition ABSOLUTELY CONVERGENT—A series $\sum_{k=1}^{\infty} a_k$ converges absolutely if $\sum_{k=1}^{\infty} |a_k|$ converges.

Note: If a series converges absolutely, it converges.

Definition CONDITIONALLY CONVERGENT—A series $\sum_{k=1}^{\infty} a_k$ converges conditionally if it converges but $\sum_{k=1}^{\infty} |a_k|$ diverges.

Note: If we have an alternating series, to show it converges conditionally, we only have to show its terms go to zero. To find out whether it is absolutely convergent, we must use some other test.

Note: There are three possibilities for an alternating series: diverge, converge conditionally, converge absolutely.

Let us look at 3 alternating series.

EXAMPLE 31 Let us look at $\sum_{k=1}^{\infty} (-1)^{k+1}/(2k+1)$. This series converges conditionally since (1) the terms go to zero, but (2) using the limit comparison test with the harmonic series, the positive series behaves as the harmonic series and diverges.

EXAMPLE 32 What about the series $\sum_{k=1}^{\infty} (-1)^{k+1}/(k^2+1)$? This series converges absolutely by comparing to the p_2 series.

EXAMPLE 33 $\sum_{k=1}^{\infty} (-1)^{k+1}k^2/(k^2+6)$ diverges since the terms don't go to 0.

Definition REGION OF CONVERGENCE—We have an infinite series whose terms are functions of x. The set of all points, x, for which the series converges is called the region of convergence.

Now let's get back to the series with x's in them. Series of this type are usually done with the RATIO TEST. This is to find the region of convergence. Then you will test both the left and right end points, THREE TESTS IN ALL.

EXAMPLE 34 $\sum_{k=1}^{\infty} x^k/k$

Using the ratio test,

$|a_{k+1}/a_k| = |x^{k+1}/(k+1)|$ divided by $|x^k/k|$

$$= \left|\frac{x^{k+1}}{k+1}\right| \times \left|\frac{k}{x^k}\right| = \left|\frac{kx}{k+1}\right| \cdot \lim_{k\to\infty}\left|\frac{kx}{k+1}\right| = |x|$$

So the region of convergence is $|x|<1$ or $-1<x<1$.

Let us test both 1 and -1 by substituting those values into the original series. $x=1$ gives us $\sum_{k=1}^{\infty} 1^k/k$ or $\sum_{k=1}^{\infty} 1/k$, the harmonic series which diverges. $x=-1$ gives us $\sum_{k=1}^{\infty} (-1)^k/k$, the alternating harmonic series, or rather the negative of the alternating harmonic series since the first term is negative. We know this converges. Therefore the region of convergence is $-1 \le x < 1$.

Important note: When you test the end points, anything is possible. Both ends could converge, both diverge, the left converge and not the right, or the right converge but not the left.

EXAMPLE 35 Let's look at $\sum^{\infty} (x-4)^k/3^k$.

This is a geometric series that converges for $|r| = \left|\frac{x-4}{3}\right| < 1$. Thus the region of convergence is $|x-4|<3$ or $1<x<7$. Test $x=1$ and substitute into the original series. We get

$$\sum_{k=1}^{\infty} \frac{(-3)^k}{3^k} \text{ or } \sum_{k=1}^{\infty} (-1)^k = -1+1-1+1\cdots$$

which diverges (Example 11). For $x=7$ we get

$$\sum_{k=1}^{\infty} \frac{3^k}{3^k} = 1+1+1+1+\cdots$$

which diverges (Example 10).

EXAMPLE 36 $\sum_{k=1}^{\infty} x^k/k!$, a nice one.

$|a_{k+1}/a_k| = |x^{k+1}/(k+1)|$ divided by $|x^k/k!|$

$$= \left|\frac{x^{k+1}}{(k+1)!}\right| \times \left|\frac{k!}{x^k}\right| = \left|\frac{xk!}{(k+1)k!}\right| = \left|\frac{x}{k+1}\right|$$

$\lim_{k\to\infty}\left|\frac{x}{k+1}\right| = 0$. This says no matter what x is, the limit will always be less than 1. The region of convergence is ALL REAL NUMBERS.

EXAMPLE 37 $\sum_{k=1}^{\infty} (k+3)!x^k$

$|a_{k+1}/a_k| = |(k+4)!x^{k+1}|$ divided by $|(k+3)!x^k|$

$$= \left|\frac{(k+4)(k+3)!x^{k+1}}{(k+3)!x^k}\right| = |(k+4)x|$$

$\lim_{k\to\infty} |(k+4)x| \to \infty$ except if $x = 0$. The region of convergence is just the point $x = 0$.

Which test to use

After finishing this book but before running off copies, I finished student-testing this section on infinite series. It became absolutely clear this page is necessary.

1. Always see if the terms go to zero first. If they don't, the series diverges. If the terms go to zero, the series at least converges conditionally if it alternates.
2. Use the integral test if the infinite series looks like an integral you have done. By this time you should have so many integrals you should be familiar with and/or sick of them.
3. Don't use the integral test if you can see an easier one or if there is a factorial symbol.
4. My favorite is the ratio test. Always try the ratio test if there is a factorial or an x in the problem. Also try the ratio test if there is something to a power of k, such as 2^k or k^k.
5. Use the limit comparison test or the comparison test if the series looks like one you know like the harmonic series, p_2 series, etc. Use the comparison if the algebra is not too bad. Use the limit comparison if the algebra looks really terrible, or even semiterrible.
6. Use the root test if there is at least 1 term with k in the exponent and no factorial in the problem.
7. If there is a series where there are a lot of messy-looking terms multiplying each other, the ratio test is probably the correct one.
8. Sometimes you may not be able to tell the terms go to zero. The ratio test may give absolute convergence or divergence immediately.
9. Practice factorial. It is new to most of you. Once again note that $(2n+1)! = (2n+1)(2n)! = (2n+1)(2n)(2n-1)!$, that is, $7! = 7(6!) = 7(6)(5!)$. Study factorial!!!!!!.

10. Most of all, do a lot of series testing. You will get better if you practice. The nice part is the problems are mostly very short.

A PREVIEW OF POWER SERIES

We would like to have a polynomial approximation of a function in the vicinity of a given point. Polynomials are very easy to work with. They can be integrated easily while many functions can't be integrated at all. Exact answers are usually not needed since we do not live in a perfect world (or do we?).

We therefore have Taylor's theorem which gives us a polynomial that approximates f(x) for every x approximately equal to a; the closer x is to a, the better the approximation for a given length. The "meat" of the theorem is a formula for the remainder, or error, when you replace the function by the polynomial. This is necessary so that you know how close your answer is.

Taylor's theorem

1. $f^{(n+1)}(x)$, [n + 1 derivatives], continuous on some interval, I, where x is in the interval.
2. a is any number in the interval I, usually its midpoint.
3. $$S_n(x) = f(a) + \frac{f'(a)(x-a)}{1!} + \frac{f''(a)(x-a)^2}{2!} + \frac{f'''(a)(x-a)^3}{3!} + \cdots + \frac{f^{(n)}(a)(x-a)^n}{n!}$$

   Note: $S_n(x)$ is the sum of the polynomial terms up to term of degree n.
4. The remainder $R_n(x) = f(x) - S_n(x)$ for all x in I.

Then there is a point w in I—w is between a and x—such that

$$R_n(x) = \frac{f^{(n+1)}(w)(x-a)^{n+1}}{(n+1)!}$$

Let us give 3 examples worked out all the way.

EXAMPLE 38 $f(x) = e^x$. $a = 0$.

Write polynomial of degree 2. Write the remainder. Find the approximate value for $e^{.2}$ and estimate the maximum error from the actual value of $e^{.2}$.

This sounds like a lot of work, but as we will see, this process for e^x (e doesn't stand for "easy" but it should) is really quite short.

$$f(x) = e^x \qquad f'(x) = e^x \qquad f''(x) = e^x \qquad f'''(x) = e^x$$

$$f(0) = f'(0) = f''(0) = 1 \qquad f'''(w) = e^w$$

$$f(x) = \underbrace{f(a) + \frac{f'(a)(x-a)}{1!} + \frac{f''(a)(x-a)^2}{2!}}_{S_2(x)} + \underbrace{\frac{f'''(w)(x-a)^3}{3!}}_{R_2(x)}$$

$$e^x = 1 + x/1 + x^2/2 + e^w x^3/6$$

Therefore $S_2(.2) = 1 + .2 + (.2)^2/2 = 1.22$ and $R_2(.2) = e^w(.2)^3/6$ where w is between 0 and .2. Because e^x is an increasing function, $e^w < e^{.2} < e^{.5} < (3)^{1/2} < 2$ (being very wasteful). Therefore, $R_2(.2) < 2(.2)^3/6 \approx .00267$, the maximum error.

This is a pretty good approximation. Remember we were really rough-estimating the error, and this is only a polynomial of degree 2.

EXAMPLE 39 Let us do the same for $\ln(1 + x)$, polynomial degree 3, $a = 0$, $x = 1$, estimate the error for $\ln 1.1$.

$$f(x) = \ln(x+1) \qquad f'(x) = (x+1)^{-1} \qquad f''(x) = (-1)(x+1)^{-2}$$

$$f'''(x) = (-1)(-2)(x+1)^{-3} \qquad f''''(x) = (-1)(-2)(-3)(x+1)^{-4}$$

$$f(0) = 0 \qquad f'(0) = 1 \qquad f''(0) = -1 \qquad f'''(0) = 2 \qquad f''''(w) = -6(w+1)^{-4}$$

$$f(x) = f(a) + f'(a)(x-a) + \frac{f''(a)(x-a)^2}{2!} + \frac{f'''(a)(x-a)^3}{3!} + \frac{f''''(w)(x-a)^4}{4!}$$

$$\ln(1+x) = 0 + x - 1x^2/2 + 2x^3/6 - 6(w+1)^{-4}x^4/4!$$

Therefore $S_3(x) = x - x^2/2 + x^3/3$ and $\ln(1.1) = .1 - (.1)^2/2 + (.1)^3/3 = .098333\cdots$. $R_3(x) = -x^4/4(w+1)^4$.

Let's estimate. $0 < w < .1$. So $1 < w + 1 < 1.1$. And soooo $1/1.1 < 1/(w+1) < 1$. $|R_3(.1)| = (.1)^4/4(w+1)^{-4} < \frac{.1^4}{4} = .000025$. Not bad!!

EXAMPLE 40 Let's do the same for $f(x) = \sin x$ except let $a = 30° = \pi/6$, with a polynomial of degree 3. $x = 32°$. Hold on to your hats, 'cause this is pretty messy.

$f(x) = \sin x \qquad f'(x) = \cos x \qquad f''(x) = -\sin x \qquad f'''(x) = -\cos x$

$f^{(4)}(x) = \sin x \qquad f(\pi/6) = \frac{1}{2} \qquad f'(\pi/6) = 3^{1/2}/2 \qquad f''(\pi/6) = \frac{1}{2}$

$f'''(\pi/6) = 3^{1/2}/2 \qquad f^{(4)}(w) = \sin w$

$$f(x) = f(a) + f'(a)(x-a) + \frac{f''(a)(x-a)^2}{2!} + \frac{f'''(a)(x-a)^3}{3!} + \frac{f^{(4)}(w)(x-a)^4}{4!}$$

$$\sin x = \frac{1}{2} + 3^{1/2}/2\,\frac{(x-\pi/6)}{1!} - \frac{\frac{1}{2}(x-\pi/6)^2}{2!} - \frac{(3^{1/2}/2)(x-\pi/6)^3}{3!} + \frac{(\sin w)(x-\pi/6)^4}{4!}$$

Now $x = 32° = 32\pi/180$. So $x - \pi/6 = \frac{32\pi}{180} - \frac{30\pi}{180} = \frac{\pi}{90} = .035$.

in 32 degrees $= \frac{1}{2} + (3^{1/2}/2)(.035) - \frac{\frac{1}{2}(.035)^2}{2} - \frac{(3^{1/2}/2)(.035)^3}{6} =$ .5299985, accuracy not guaranteed since $\pi/90$ should be more places.

Actually I couldn't bear this inaccuracy. So $\pi/90$ is .0349066 on my calculator, and if I hit the right buttons, $\sin 32° = .5299195$ (and my calculator is OK). The remainder is $(\sin w)(.0349066)^4/4!$ We know $\sin w$ is less than 1, so the remainder is less than $(.0349066)^4/4! = 6.1861288 \times 10^{-8}$.

Even with these limited examples, we see different series get more accurate results with the same number of terms. This can be studied in great detail.

Also it is very convenient to know certain power series by heart. We will list the most important together with region of convergence.

e^x all reals $1 + x + x^2/2! + x^3/3! + \cdots + x^n/n! + \cdots \quad n = 0,1,2,3,\ldots$

$\sin x$ all reals $x/1! - x^3/3! + x^5/5! - x^7/7! \ldots (-1)^k x^{2k+1}/(2k+1)! \ldots$
$k = 0,1,2,3,\ldots$

$\cos x$ all reals $1 - x^2/2! + x^4/4! - x^6/6! \cdots (-1)^k x^{2k}/(2k)! \ldots$
$k = 0,1,2,3,\ldots$

$\ln(x+1)$ $-1 < x \le 1$ $x - x^2/2 + x^3/3 - x^4/4 \ldots (-1)^{k+1} x^k/k \ldots$
$k = 1,2,3,4,\ldots$

Here are some more . . .

$1/(1-x)$ $\quad -1<x<1$ $\quad 1+x+x^2+x^3+\cdots+x^k+\cdots$
$k=0,1,2,3,\ldots$

$1/(x+1)$ $\quad -1<x<1$ $\quad 1-x+x^2-x^3+\cdots(-1)^k x^k$ $\quad k=0,1,2,3,\ldots$

Binomial $f(x)=(1+x)^p$ $\quad -1\le x<1$ $\quad 1+px+p(p-1)x^2/2!+$
$p(p-1)(p-2)x^3/3!+\cdots+p(p-1)\cdots[p-(n-1)]x^n/n!\ldots$
$n=0,1,2,3,\ldots$

Finally there are theorems found in many books which give the conditions under which you can add, subtract, multiply, divide, differentiate, integrate infinite series. We can amaze ourselves by the number of functions we can approximate.

EXAMPLE 41 Find the infinite series for cosh x.

$$\cosh x=(e^x+e^{-x})/2$$
$$e^x=1+x+x^2/2!+x^3/3!+x^4/4!+\cdots$$
$$e^{-x}=1-x+x^2/2!-x^3/3!+x^4/4!-\cdots \quad \textbf{Substitute } -x \textbf{ for } x$$
$$\cosh x=1+x^2/2!+x^4/4!+x^6/6!+\cdots \quad \textbf{Add and divide by 2}$$

Pretty neat, eh?! More to come.

EXAMPLE 42 $\int_0^1 e^{-x^2}\,dx$, 4 terms

$$e^x=1+x+x^2/2!+x^3/3!$$
$$e^{-x^2}=1-x^2+x^4/2-x^6/6$$
$$\int_0^1 e^{-x^2}\,dx=x-x^3/3+x^5/10-x^7/42$$
$$=1-1/3+1/10-1/42=78/105$$

Not too shabby. More to come.

EXAMPLE 43 The series for $x/(1+x)^2$

$$1/(1+x)=1-x+x^2-x^3+x^4+\cdots \quad \textbf{Differentiating we get}$$
$$-1/(1+x)^2=-1+2x-3x^2+4x^3-5x^4+\cdots \quad \textbf{Multiply by } -x\textbf{; our result is}$$
$$x/(1+x)^2=x-2x^2+3x^3-4x^4+5x^5-\cdots$$

When mathematicians do things like this, you tend to believe that mathematics can do everything and anything. However this is not true. However the best is yet to come!!!!!!

We can derive every property of the sine and cosine using infinite series and NEVER, NEVER mentioning triangles or angles. Amazing, huh?! Given $\sin x = x - x^3/3! + x^5/5! \ldots$ and $\cos x = 1 - x^2/2! + x^4/4! - x^6/6! \ldots$. How about $\cos 2x$? $\cos 2x = 1 - 2x^2 + (2/3)x^4 - (4/45)x^6 \ldots$ [2x for x in cos x].

How about $\sin^2 x + \cos^2 x = 1$?

$$\begin{aligned}\sin^2 x + \cos^2 x &= (x - x^3/6 + x^5/120 \cdots)(x - x^3/6 + x^5/120 \cdots)\\ &\quad + (1 - x^2/2 + x^4/24 - x^6/720 \cdots)\\ &\quad \times (1 - x^2/2 + x^4/24 - x^6/720 \cdots)\\ &= x^2 - x^4/3 + 2x^6/45 \cdots + 1 - x^2 + x^4/3 - 2x^6/45 \cdots\\ &= 1 + 0 + 0 + 0 \cdots = 1\end{aligned}$$

How about tan x? Well we would like $\tan x = \sin x/\cos x$.

$$1 - x^2/2 + x^4/24 - x^6/72 \cdots \overline{\big)\, x - x^3/6 + x^5/120 - x^7/540 \cdots}^{\;x + x^3/3 + 2x^5/15 + 17x^7/315 \cdots} = \tan x$$

How about derivatives? If $f(x) = \sin x$, we want $f'(x) = \cos x$.

$$\begin{aligned}\sin x &= x - x^3/6 + x^5/120 - x^7/5040 \cdots\\ (\sin x)' &= 1 - 3x^2/6 + 5x^4/120 - 7x^6/5040 \cdots\\ &= 1 - x^2/2 + x^4/24 - x^6/720 \cdots = \cos x\end{aligned}$$

We could, of course, integrate cos x term by term and get sin x. We could go on and on getting every property of the sine and cosine and all other trig functions totally without angles or triangles. The beauty of these power series is that they are limits of polynomials and are easy to deal with. Yet there are things that are even more powerful in mathematics. But they must wait for another book.

ACKNOWLEDGMENTS

I have many people to thank.

I would like to thank my wife Marlene who makes life worth living.

I thank the two most wonderful children in the world, Sheryl and Eric for being themselves.

I would like to thank my brother Jerry for all his encouragement and for arranging to have my non-professional editions printed.

I would like to thank Bernice Rothstein of the City College of New York and Sy Solomon at Middlesex County Community College book stores for allowing my books to be sold in their book stores and for their kindness and encouragement.

I would like to thank Dr. Robert Urbanski, chairman of the math department at Middlesex, first for his encouragement, and secondly for recommending my books to his students because the students found the books valuable.

I thank Bill Summers of the CCNY audio-visual department for his help on this and other endeavors.

Next I would like to thank the backbone of three schools, their secretaries: Hazel Spencer of Miami of Ohio, Libby Alam and Efua Tongé of the City College of New York, and Sharon Nelson of Rutgers.

I would like to thank Marty Levine of MARKET SOURCE for first presenting my books to McGraw-Hill.

I would like to thank McGraw-Hill, especially John Carleo, John Aliano, David Beckwith and Pat Koch. Hopefully this will be the beginning of a long and mutually beneficial relationship.

I would also like to thank my parents, Lee and Cele, who saw the beginnings of these books but did not live to see their publication.

Lastly I would like to thank three people who helped keep my spirits up when things looked very bleak: a great friend, Gary Pitkofsky, another terrific friend and fellow lecturer David Schwinger, and my sharer of dreams, my cousin, Keith Ellis, who also did not live to see my books published.

INDEX

MY NOTES

MY NOTES

MY NOTES

MY NOTES